D0796783

HERBS BEING DUG UP AND MADE INTO MEDICINES UNDER THE DIRECTION OF A SAGE

From a 12th century copy of the *Herbarium of Apuleius,* now in the Library of Eton College

THE OLD ENGLISH HERBALS

BY

ELEANOUR SINCLAIR ROHDE

AUTHOR OF "A GARDEN OF HERBS"

Illustration of the "lilie" from the Saxon translation of the *Herbarium of Apuleius*

DOVER PUBLICATIONS, INC.

NEW YORK

Published in Canada by General Publishing Company, Ltd., 30 Lesmill Road, Don Mills, Toronto, Ontario.

Published in the United Kingdom by Constable and Company, Ltd., 10 Orange Street, London WC 2.

This Dover edition, first published in 1971, is an unabridged and unaltered republication of the work originally published by Longmans, Green and Co. in 1922.

The frontispiece was reproduced in full color in the original edition.

International Standard Book Number: 0-486-21106-1
Library of Congress Catalog Card Number: 75-166433

Manufactured in the United States of America
Dover Publications, Inc.
180 Varick Street
New York, N.Y. 10014

TO

MY BROTHER

"The Lely is an herbe wyth a whyte floure. And though the levys of the floure be whyte: yet wythin shyneth the lykenesse of golde."—Bartholomæus Anglicus (*circ.* 1260).

PREFACE

THE writing of this book on that fascinating and somewhat neglected[1] branch of garden literature—the old English Herbals—has been a labour of love, but it could not have been done without all the kind help I have had. My grateful thanks are due to the authorities at the British Museum, to Professor Burkitt of Cambridge, and very specially to Mr. J. B. Capper for invaluable help. I am indebted to Dr. James, the Provost of Eton, for his kind permission to reproduce an illustration from a twelfth-century MS. in the Library of Eton College for the frontispiece. I find it difficult to express either my indebtedness or my gratitude to Dr. and Mrs. Charles Singer, the former for all his help and the latter for her generous permission to make use of her valuable bibliography of early scientific manuscripts. I am further indebted to Dr. Charles Singer for reading the chapter on the Anglo-Saxon herbals in proof. For their kind courtesy in answering my inquiries concerning the MS. herbals in the libraries of their respective cathedrals, I offer my grateful thanks to the Deans of Lincoln and Gloucester Cathedrals, and to the Rev. J. N. Needham for information concerning the herbals in the library of Durham Cathedral; to the librarians of the following colleges—All Souls' College, Oxford; Balliol College, Oxford; Corpus Christi College, Oxford; Corpus Christi College, Cambridge; Emmanuel College, Cambridge; Gonville and Caius College, Cambridge; Magdalene College, Cambridge; Peterhouse, Cambridge; Jesus College, Cambridge; St. John's College, Oxford, and Trinity College, Cambridge;

[1] It is a remarkable fact that even the eleventh edition of the omniscient *Encyclopædia Britannica* has no article on Herbals.

to the librarians of Durham University, Trinity College, Dublin, the Royal Irish Academy, and the National Library of Wales; to the Honble. Lady Cecil for information respecting MSS. in the library of the late Lord Amherst of Hackney; and to the following owners of private libraries—the Marquis of Bath, Lord Leconfield, Lord Clifden, Mr. T. Fitzroy Fenwick of Cheltenham, and Mr. Wynne of Peniarth, Merioneth. For information respecting incunabula herbals in American libraries I am indebted to Dr. Arnold Klebs and to Mr. Green of the Missouri Botanical Garden, St. Louis.

No pains have been spared to make the bibliographies as complete as possible, but I should be glad to be told of any errors or omissions. There are certain editions of *Banckes's Herbal* and *The Grete Herball* mentioned by authorities such as Ames, Hazlitt, etc., of which no copies can now be found in the chief British libraries (see p. 204 *et seq.*). If any copies of these editions are in private libraries I should be grateful to hear of them. The rarest printed herbal is "*Arbolayre contenāt la qualitey et vertus proprietiez des herbes gōmes et simēces extraite de plusiers tratiers de medicine com̄ent davicene de rasis de constatin de ysaac et plateaire selon le con̄u usaige bien correct.*" (Supposed to have been printed by M. Husz at Lyons.) It is believed that there are only two copies of this book now extant. One is in the Bibliothèque Nationale, Paris; the other was sold in London, March 23, 1898, but I have been unable to discover who is the present owner. For this or any other information I should be most grateful.

ELEANOUR SINCLAIR ROHDE.

CONTENTS

CHAPTER I

CHAPTER II

LIST OF ILLUSTRATIONS

THE OLD ENGLISH HERBALS

CHAPTER I

THE ANGLO-SAXON HERBALS

"Everything possible to be believ'd is an image of truth."
WILLIAM BLAKE.

THERE is a certain pathos attached to the fragments from any great wreck, and in studying the few Saxon manuscripts, treating of herbs, which have survived to our day, we find their primary fascination not so much in their beauty and interest as in the visions they conjure up of those still older manuscripts which perished during the terrible Danish invasions. That books on herbs were studied in England as early as the eighth century is certain, for we know that Boniface, "the Apostle of the Saxons," received letters from England asking him for books on simples and complaining that it was difficult to obtain the foreign herbs mentioned in those we already possessed.[1] But of these manuscripts none have survived, the oldest we possess being of the tenth century, and for our knowledge of Anglo-Saxon plant lore we look chiefly to those four important manuscripts—the *Leech Book of Bald*, the *Lacnunga* and the Saxon translations of the *Herbarium of Apuleius* and the so-called Περὶ Διδαξέων.

Apart from their intrinsic fascination, there are certain considerations which give these manuscripts a peculiar importance. Herb lore and folk medicine lag not years, but centuries,

[1] Nec non et si quos sæcularis scientiæ libros nobis ignotos adepturi sitis, ut sunt de medicinalibus, quorum copia est aliqua apud nos, sed tamen segmenta ultra marina quæ in eis scripta comperimus, ignota nobis sunt et difficilia ad adipiscendum.—Bonifac., *Epistolæ,* p. 102.

behind the knowledge of their own day. Within living memory our peasants were using, and in the most remote parts of these islands they use still, the herbal and other remedies of our Saxon ancestors. They even use curiously similar charms. The herb lore recorded in these manuscripts is the herb lore, not of the century in which they were written, but of the dim past ages pictured in the oldest parts of *Widsith* and *Beowulf.* To the student of English plant lore, the *Herbarium of Apuleius* and the Περὶ Διδαξέων are less interesting because they are translations, but the more one studies the original Saxon writings on herbs and their uses, the more one realises that, just as in *Beowulf* there are suggestions and traces of an age far older than that in which the poem was written, so in these manuscripts are embedded beliefs which carry us back to the dawn of history. It is this which gives this plant lore its supreme interest. It is almost overwhelming to recognise that possibly we have here fragments of the plant lore of our ancestors who lived when Attila's hordes were devastating Europe, and that in the charms and ceremonies connected with the picking and administering of herbs we are carried back to forms of religion so ancient that, compared to it, the worship of Woden is modern. Further, it is only in these manuscripts that we find this herb lore, for in the whole range of Saxon literature outside them there is remarkably little mention of plant life. The great world of nature, it is true, is ever present; the ocean is the background of the action in both *Beowulf* and *Cynewulf,* and the sound of the wind and the sea is in every line. One is conscious of vast trackless wastes of heath and moor, of impenetrable forests and terror-infested bogs; but of the details of plant life there is scarcely a word. In these manuscripts alone do we find what plant life meant to our ancestors, and, as with all primitive nations, their belief in the mystery of herbs is almost past our civilised understanding. Their plant lore, hoary with age, is redolent of a time when the tribes were still wandering on the mainland of Europe, and in these first records of this plant lore there is the breath of mighty forests, of marsh lands and of Nature in her wildest. We are swept back to an epoch when man fought with Nature, wresting from her the land, and when the unseen powers of evil resented this conquest of their domains. To the early Saxons those unseen powers were an everyday reality. A supernatural terror brooded over the trackless heaths, the dark mere pools were inhabited by the water elves. In the wreathing mists and driving storms of snow and hail they saw the uncouth " moor gangers," " the muckle mark steppers who hold the moors," or the stalking fiends of the lonely places, creatures whose baleful eyes shone like flames through the mist. To this day some of our place names in the more remote parts of these islands recall the memory of those evil terrors. In these manuscripts we are again in an atmosphere of eotens and trolls, there are traces of even older terrors, when the first Teuton settlers in Europe struggled with the aborigines who lived in caves, hints as elusive as the phantom heroes in the Saxon poems, and as unforgettable.

Still more remarkable is the fact that beneath the superstructure of Christian rites to be used when the herbs were being picked or administered we find traces not merely of the ancient heathen religion, but of a religion older than that of Woden. It has been emphasised by our most eminent authorities that in very early times our ancestors had but few chief gods, and it is a remarkable fact that there is no mention whatever of Woden in the whole range of Saxon literature before the time of Alfred. In those earlier centuries they seem to have worshipped a personification of Heaven, and Earth, the wife of Heaven, and the Son, whom after ages called Thor. There were also Nature deities, Hrede, the personification of the brightness of Summer, and Eostra, the radiant creature of the Dawn. It will be remembered that it was the worship, not of Balder, but of Eostra, which the Christian missionaries found so deeply imbedded that they adopted her name and transferred it to Easter. For this we have the authority of Bede. Separate from these beneficent powers were the destroying and harmful powers of Nature—darkness, storm, frost and the deadly vapours of moorland and fen, personified in the giants, the ogres, the furious witches that rode the winds and waves; in fact, the whole horde of demons of sea and land and sky. It is the traces of these most ancient forms of religion which give to the manuscripts their strongest fascination.

Many of us miss all that is most worth learning in old books through regarding anything in them that is unfamiliar as merely quaint, if not ridiculous. This attitude seals a book as effectually and as permanently as it seals a sensitive human being. There is only one way of understanding these old writers, and that is to forget ourselves entirely and to try to look at the world of nature as they did. It is not "much learning" that is required, but sympathy and imagination. In the case of these Saxon manuscripts we are repaid a thousandfold; for they transport us to an age far older than our own, and yet in some ways so young that we have lost its magic key. For we learn not only of herbs and the endless uses our forefathers made of them, but, if we try to read them with understanding, these books open for us a magic casement through which we look upon the past bathed in a glamour of romance. Our Saxon ancestors may have been a rude and hardy race, but they did not live in an age of materialism as we do. In their writings on herbs and their uses we see "as through a glass darkly" a time when grown men believed in elves and goblins as naturally as they believed in trees, an age when it was the belief of every-day folk that the air was peopled with unseen powers of evil against whose machinations definite remedies must be applied. They believed, as indeed the people of all ancient civilisations have believed, that natural forces and natural objects were endued with mysterious powers whom it was necessary to propitiate by special prayers. Not only the stars of heaven, but springs of water and the simple wayside herbs, were to them directly associated with unseen beings. There are times when one is reminded forcibly of that worship of Demeter, " nearer to the Earth which some have thought they could discern behind the definitely national mythology of Homer." They believed that the sick could be cured by conjurations and charms, as firmly as we believe to-day in curing them by suggestion—is there any real difference between these methods? —and when one reads the charms which they used in administering their herbs one cannot help wondering whether these were handed down traditionally from the Sumerians, those ancient inhabitants of Mesopotamia who five thousand years before Christ used charms for curing the sick which have now been partially deciphered from the cuneiform inscriptions. But before studying the plant lore therein contained, it may be as well to take a preliminary survey of the four most important manuscripts.

The oldest Saxon book dealing with the virtues of herbs which we possess is the *Leech Book of Bald,* dating from about A.D. 900–950. Unlike some other MS. herbals of which only a few tattered pages remain, this perfect specimen of Saxon work has nothing fragile about it. The vellum is as strong and in as good condition as when it first lay clean and untouched under the hand of the scribe—Cild by name—who penned it with such skill and loving care. One's imagination runs riot when one handles this beautiful book, now over a thousand years old, and wonders who were its successive owners and how it has survived the wars and other destructive agencies through all these centuries. But we only know that, at least for a time, it was sheltered in that most romantic of all English monasteries, Glastonbury.[1] This Saxon manuscript has a dignity which is unique, for it is the oldest existing leech book written in the vernacular. In a lecture delivered before the Royal College of

[1] A catalogue of the books of that foundation cited by Wanley (Hickes, *Thesaur.* Vol. II. Præf. ad Catalogum) contains the entry " Medicinale Anglicum," and the MS. described above has on a fly-leaf the now almost illegible inscription " Medicinale Anglicum." There is unfortunately no record as to the books which, on the dissolution of the monasteries, may possibly have found their way from Glastonbury to the royal library.

Physicians in 1903, Dr. J. F. Payne commented on the remarkable fact that the Anglo-Saxons had a much wider knowledge of herbs than the doctors of Salerno, the oldest school of medicine and oldest university in Europe. "No treatise," he said, "of the School of Salerno contemporaneous with the *Leech Book of Bald* is known, so that the Anglo-Saxons had the credit of priority. Their Leech Book was the first medical treatise written in Western Europe which can be said to belong to modern history, that is, which was produced after the decadence and decline of the classical medicine, which belongs to ancient history. . . . It seems fair to regard it [the Leech Book], in a sense, as the embryo of modern English medicine, and at all events the earliest medical treatise produced by any of the modern nations of Europe." The Anglo-Saxons created a vernacular literature to which the continental nations at that time could show no parallel, and in the branch of literature connected with medicine, in those days based on a knowledge of herbs (when it was not magic), their position was unique. Moreover, the fact that the Leech Book was written in the vernacular is in itself remarkable, for it points to the existence of a class of men who were not Latin scholars and yet were able and willing to read books. The Leech Book belongs to the literary period commonly known as the school of Alfred. It was probably written shortly after Alfred's death, but it is more than probable that it is a copy of a much older manuscript, for what is known as the third book of the Leech Book is evidently a shorter and older work incorporated by the scribe when he had finished the Leech Book proper.

The book itself was written under the direction of one Bald, who, if he were not a personal friend of King Alfred's, had at any rate access to the king's correspondence; for one chapter consists of prescriptions sent by Helias, Patriarch of Jerusalem, to the king.[1] We learn the names of the first owner and scribe

[1] This chapter consists of prescriptions containing drugs such as a resident in Syria would recommend. It is interesting to find this illustration of Asser's

from lines in Latin verse at the end of the second part of the MS.

> " Bald is the owner of this book, which he ordered Cild to write,
> Earnestly I pray here all men, in the name of Christ,
> That no treacherous person take this book from me,
> Neither by force nor by theft nor by any false statement.
> Why? Because the richest treasure is not so dear to me
> As my dear books which the Grace of Christ attends."

The book consists of 109 leaves and is written in a large, bold hand and one or two of the initial letters are very faintly illuminated. The writing is an exceptionally fine specimen of Saxon penmanship. On many of the pages there are mysterious marks, but it is impossible to conjecture their meaning. It has been suggested that they point to the sources from which the book was compiled and were inserted by the original owner.

The *Leech Book of Bald* was evidently the manual of a Saxon doctor, and he refers to two other doctors—Dun and Oxa by name—who had given him prescriptions. The position of the leech in those days must have been very trying, for he was subjected to the obviously unfair competition of the higher clergy, many of whom enjoyed a reputation for working miraculous cures.[1] The leech being so inferior in position, it is not surprising that his medical knowledge did not advance on

---

statement, that he had seen and read the letters which the Patriarch of Jerusalem sent with presents to the king. From Asser also we learn that King Alfred kept a book in which he himself entered " little flowers culled on every side from all sorts of masters." " Flosculos undecunque collectos a quibus libet magistris et in corpore unius libelli mixtim quamvis sicut tunc suppetebat redigere."—ASSER, p. 57.

[1] The stories of miraculous cures by famous Anglo-Saxon bishops and abbots are for the most part too well known to be worth quoting, but the unfair treatment of the leech is perhaps nowhere more clearly shown than in Bede's tale of St. John of Beverley curing a boy with a diseased head. Although the leech effected the cure, the success was attributed to the bishop's benediction, and the story ends, " the youth became of a clear countenance, ready in speech and with hair beautifully wavy."

scientific lines. He relied on the old heathen superstitions, probably from an instinctive feeling that in pagan religion, combined with the herb lore which had been handed down through the ages, the mass of the people had a deep-rooted faith. Nothing is more obvious in the Leech Book than the fact that the virtues ascribed to the different herbs are based not on the personal knowledge of the writer, but on the old herb lore. This gives the Leech Book its special fascination; for it is the oldest surviving manuscript in which we can learn the herb lore of our ancestors, handed down to them from what dim past ages we can only surmise. We have, therefore, to bear in mind that what may strike our modern minds as quaint, or even grotesque, is in the majority of instances a distorted form of lore which doubtless suffered many changes during the early centuries of our era. Nearly all that is most fascinating in the Leech Book is of very ancient Indo-Germanic or Eastern origin, but one cannot help wondering how much the Saxons incorporated of the herb lore of the ancient Britons. Does not Pliny tell us that the Britons gathered herbs with such striking ceremonies that it would seem as though the Britons had taught them to the Persians?

One cannot read Bald's manuscript without being struck by his remarkable knowledge of native plants and garden herbs. We are inferior to our continental neighbours in so many arts that it is pleasant to find that in the ancient art of gardening and in their knowledge of herbs our Saxon forefathers excelled. It has been pointed out by eminent authorities that the Anglo-Saxons had names for, and used, a far larger number of plants than the continental nations. In the *Herbarium of Apuleius*, including the additions from Dioscorides, only 185 plants are mentioned, and this was one of the standard works of the early Middle Ages. In the *Herbarius* of 1484, the earliest herbal printed in Germany, only 150 plants are recorded, and in the German *Herbarius* of 1485 there are 380. But from various sources it has been computed that the Anglo-Saxons had names for, and used, at least 500 plants.[1] One feels instinctively that the love of flowers and gardens was as deep-rooted in our ancestors as it is in our nation to-day, and though we do not know exactly what they grew in their gardens—which they called wyrtȝerd (literally, herb-yard)—we do know that the marigolds, sunflowers, peonies, violets and gilly-flowers which make the cottage gardens of England so gay and full of colour to-day were also the commonest plants in the Saxon gardens. Fashions in large gardens have changed throughout the centuries, and there are stately gardens in this country famed the world over. But in regard to our cottage gardens we are staunchly conservative, and it is assuredly the cottage garden which is characteristically English. Incidentally, one cannot help regretting that so many of our old Saxon plant names have fallen into disuse. "Waybroad," for instance, is much more descriptive than "plantain," which is misleading.[2] "Maythen" also is surely preferable to "camomile," and "wergulu" is more characteristic of that fierce weed than "nettle." Those of us who are gardeners will certainly agree that "unfortraedde" is the right name for knotweed. And is not "joy of the ground" a delightful name for periwinkle?

The oldest illustrated herbal which has come down to us from Saxon times is the translation of the Latin *Herbarium Apuleii Platonici*.[3] The original Latin work is believed to date

[1] A small but striking instance of Saxon knowledge, or rather close observation, of plants is to be found in the following description of wolf's teazle in the *Herbarium of Apuleius*:—"This wort hath leaves reversed and thorny and it hath in its midst a round and thorny knob, and that is brown-headed in the blossoms and hath white seed and a white and very fragrant root." The word "reversed" is not in the original and was therefore added by the Saxon translator, who had observed the fact that all the thistle tribe protect their leaves by thorns pointing backwards as well as forwards.

[2] It is interesting to remember that even as late as the sixteenth century plantain was called "waybroad." See *Turner's Herbal*.

[3] There are numerous Latin MSS. of this book, chiefly in Italian libraries, several being in the Laurentian Library at Florence. The book was first printed at Rome, probably soon after 1480, by Joh. Philippus de Lignamine, who was also the editor. De Lignamine, who was physician to Pope Sixtus IV.,

from the fifth century, though no copy so ancient as this is in existence now. The name Apuleius Platonicus is possibly fictitious and nothing is known of the writer, who was, of course, distinct from Apuleius Madaurensis, the author of the *Golden Ass*. The Saxon translation of this herbal (now in the British Museum) is supposed to date from A.D. 1000–1050, and belongs to the school of Ælfric of Canterbury. The frontispiece is a coloured picture in which Plato is represented holding a large volume which is being given him by Æsculapius and the Centaur, and on the other side of the page is a blue circle spotted with white and red, within which is the name of the book: "Herbarium Apuleii Platonici quod accepit ab Escolapio et Chirone centauro magistro Achillis." The book consists of 132 chapters, in each of which a herb is described, and there are accompanying illustrations of the herbs. Throughout the book there are also remarkable pictures of snakes, scorpions and unknown winged creatures. It has been pointed out that the figures of herbs are obviously not from the original plants, but are copied from older figures, and these from others older still, and one wonders what the original pictures were like. It is interesting to think that perhaps the illustrations in this Saxon herbal are directly descended, so to speak, from the drawings of Cratevas,[1] Dionysius or Metrodorus, of whom Pliny tells us "They drew the likeness of herbs and wrote under them their effects." The picture of the lily is very attractive in spite of the fact that the flowers are painted pale blue. The stamens in

---

says that he found this MS. in the library of the monastery of Monte Cassino. In the first impression the book is dedicated to Cardinal de Gonzaga; in the second impression to Cardinal de Ruvere. (The copy in the British Museum is of the second impression.) In this small quarto volume the illustrations are rough cuts. It is interesting to remember that these are the earliest known printed figures of plants. The printed text contains a large number of Greek and Latin synonyms which do not appear in the Saxon translation. Subsequent editions were printed in 1528 (Paris) and in the Aldine Collection of Latin medical writers, 1547 (Venice).

[1] Cratevas is said to have lived in the first century B.C. Pliny, Dioscorides and Galen all quote him.

AESCULAPIUS PLATO AND A CENTAUR

From the Saxon translation of the *Herbarium of Apuleius* (Cott. Vit., C. 3, folio 19*a*)

the figure stand out beyond the petals and look like rays of light, with a general effect that is curiously pleasing. One of the most interesting figures is that of the mandrake (painted in a deep madder), which embodies the old legend that it was death to dig up the root, and that therefore a dog was tied to a rope and made to drag it up. It is the opinion of some authorities that these figures show the influence of the school represented by the two splendid Vienna manuscripts of Dioscorides dating from the fifth and seventh centuries. There is no definite evidence of this, and though the illustrations in the Saxon manuscript show the influence of the classical tradition, they are poor compared with those in the Vienna manuscript. To some extent at least the drawings in this herbal must necessarily have been copies, for many of the plants are species unknown in this country.

The Saxon translation of the Περὶ Διδαξέων (Harl. 6258) is a thin volume badly mutilated in parts. Herr Max Löwenbeck [1] has shown that this is in part translated from a treatise by an eleventh-century writer, Petrocellus or Petronius, of the School of Salerno—the original treatise being entitled *Practica Petrocelli Salernitani.*[2] As has been pointed out by many eminent authorities, the School of Salerno, being a survival of Greek medicine, was uncontaminated by superstitious medicine. Consequently there are striking differences between this and the other Saxon manuscripts. The large majority of the herbs mentioned are those of Southern Europe, and the pharmacy is very simple compared with the number of herbs in prescriptions of native origin. As Dr. J. F. Payne [3] has pointed out, Herr Löwenbeck's important discovery does not account for the whole of the English book. The order of the chapters differs from that of the Salernitan writer; there are passages not to be found in the *Practica*, and in some places the English text gives

[1] Erlanger, *Beiträge zur englischen Philologie,* No. XII. (περὶ διδαξέων), eine Sammlung von Rezepten in englischer Sprache.

[2] Printed by De Renzi in *Collectio Salernitana,* Vol. IV. (Naples, 1856).

[3] *English Medicine in the Anglo-Saxon Times.*

a fuller reading. It is fairly evident that the Saxon treatise is at least in part indebted to the *Passionarius* by Gariopontus, another Salernitan writer of the same period.

The *Lacnunga* (Harl. 585), an original work, and one of the oldest and most interesting manuscripts, is a small, thick volume without any illustrations. Some of the letters are illuminated and some are rudely ornamented. At the top of the first page there is the inscription " Liber Humfredi Wanley," and it is interesting, therefore, to realise that the British Museum owes this treasure to the zealous antiquarian whose efforts during the closing years of the seventeenth and early years of the eighteenth century rescued so many valuable Saxon and other MSS. from oblivion.[1]

To the student of folk lore and folk custom these sources of herb lore are of remarkable interest for the light they throw on the beliefs and customs of humble everyday people in Anglo-Saxon times. Of kings and warriors, of bards and of great ladies we can read in other Saxon literature, and all so vividly that we see their halls, the long hearths on which the fires were piled, the openings in the roof through which the smoke passed. We see the men with their " byrnies " of ring mail, their crested

[1] On the preceding blank page there is an inscription in late seventeenth-century handwriting—

" This boucke with letters is wr [remainder of word illegible]
Of it you cane no languige make.
Ba C.
A happie end if thou dehre [dare] to make
Remember still thyn owne esstate,
If thou desire in Christ to die
Thenn well to lead thy lif applie
barbara crokker."

It is at least probable that Wanley, who at this period was collecting Anglo-Saxon manuscripts for George Hickes, secured this MS. from " barbara crokker." Her naive avowal of her inability to read the MS. suggests that she probably had no idea of the value of the book, and when one remembers Wanley's reputation for driving shrewd bargains one cannot help wondering what he paid for this treasure. Those must have been halcyon days for collectors, when a man who had been an assistant in the Bodleian Library with a salary of £12 a year could buy Saxon manuscripts !

helmets, their leather-covered shields and deadly short swords. We see them and their womenkind wearing golden ornaments at their feasts, the tables laden with boars' flesh and venison and chased cups of ale and mead. We see these same halls at night with the men sleeping, their "byrnies" and helmets hanging near them, and in the dim light we can make out also the trophies of the chase hanging on the walls. We read of their mighty deeds, and we know at least something of the ideals and the thoughts of their great men and heroes. But what of that vast number of the human kind who were always in the background? What of the hewers of wood and drawers of water, the swineherds, the shepherds, the carpenters, the hedgers and cobblers? Is it not wonderful to think that in these manuscripts we can learn, at least to some extent, what plant life meant to these everyday folk? And even in these days to understand what plant life means to the true countryman is to get into very close touch with him. Not only has suburban life separated the great concentrated masses of our people from their birthright of meadows, fields and woods; of Nature, in her untamed splendour and mystery, most of them have never had so much as a momentary glimpse. But in Saxon times even the towns were not far from the unreclaimed marshes and forests, and to the peasant in those days they were full not only of seen, but also of unseen perils. There was probably not a Saxon child who did not know something of the awe of waste places and impenetrable forests. Even the hamlets lay on the very edge of forests and moors, and to the peasant these were haunted by giant, elf and monster, as in the more inaccessible parts of these islands they are haunted still to those who retain something of primitive imagination. And when we study the plant lore of these people we realise that prince and peasant alike used the simple but mysterious herbs not only to cure them of both physical and mental ills, but to guard them from these unseen monsters. Of the reverence they paid to herbs we begin to have some dim apprehension when we read of the ceremonies connected with the picking and administering of them.

But, first, what can we learn of the beliefs as to the origin of disease? Concerning this the great bulk of the folk lore in these manuscripts is apparently of native Teutonic origin, or rather it would be more correct to speak of its origin as Indo-Germanic; for the same doctrines are to be found among all Indo-Germanic peoples, and even in the Vedas, notably the Atharva Veda. Of these beliefs, the doctrine of the "elf-shot" occupies a large space, the longest chapter in the third book of the *Leech Book of Bald* being entirely "against elf-disease." We know from their literature that to our Saxon ancestors waste places of moor and forest and marshes were the resort of a host of supernatural creatures at enmity with mankind. In the *Leech Book of Bald* disease is largely ascribed to these elves, whose shafts produced illness in their victims. We read of beorg-ælfen, dun-ælfen, muntælfen. But our modern word "elf" feebly represents these creatures, who were more akin to the "mark-stalkers," to the creatures of darkness with loathsome eyes, rather than to the fairies with whom we now associate the name. For the most part these elves of ancient times were joyless impersonations and creatures not of sun but of darkness and winter. In the gloom and solitude of the forest, "where the bitter wormwood stood pale grey" and where "the hoar stones lay thick," the black, giant elves had their dwelling. They claimed the forest for their own and hated man because bit by bit he was wresting the forest from them. Yet they made for man those mystic swords of superhuman workmanship engraved with magic runes and dipped when red hot in blood or in a broth of poisonous herbs and twigs. We do not understand, we can only ask, why did they make them? What is the meaning of the myth? The water elves recall the sea monsters who attended Grendel's dam, impersonations of the fury of the waves, akin to Hnikarr, and again other water elves of the cavernous bed of ocean, primeval deadly creatures, inhabiting alike the sea and the desolate fens, "where the elk-sedge waxed in the water." If some were akin to the Formori of the baleful fogs in Irish mythic history and the Mallt-y-nos, those she-demons of marshy lands immortalised by the Welsh bards, creatures huge and uncouth "with grey and glaring eyes," there were others who exceeded in beauty anything human. When Cædmon wrote of the beauty of Sarah, he described her as "sheen as an elf." With the passing of the centuries we have well-nigh forgotten the black elves, though they are still realities to the Highlander and too real for him to speak of them. But have we not the descendants of the sheen bright elves in the works of Shakespeare, Milton and Shelley? One feels very sure that our Saxon ancestors would have understood that glittering elf Ariel as few of us are capable of understanding him. He is the old English bright elf. Did not Prospero subdue him with magic, as our ancestors used magic songs in administering herbs "to quell the elf"? Here is one such song from the *Leech Book of Bald,* and at the end a conjuration to bury the elf in the earth.

"I have wreathed round the wounds
The best of healing wreaths
That the baneful sores may
Neither burn nor burst,
Nor find their way further,
Nor turn foul and fallow,
Nor thump and throle on,
Nor be wicked wounds,
Nor dig deeply down;
But he himself may hold
In a way to health.
Let it ache thee no more
Than ear in Earth acheth.

Sing also this many times, 'May earth bear on thee with all her might and main.' "—*Leech Book of Bald,* III. 63.

This was for one "in the water elf disease," and we read that a person so afflicted would have livid nails and tearful eyes, and would look downwards. Amongst the herbs to be administered when the charm was sung over him were a yew-berry, lupin, helenium, marsh mallow, dock elder, wormwood and strawberry leaves.

Goblins and nightmare were regarded as at least akin to elves, and we find the same herbs were to be used against them, betony being of peculiar efficacy against "monstrous nocturnal visions and against frightful visions and dreams."[1] The malicious elves did not confine their attacks to human beings; references to elf-shot cattle are numerous. I quote the following from the chapter "against elf disease."

"For that ilk [*i. e.* for one who is elf-shot].

"Go on Thursday evening when the sun is set where thou knowest that helenium stands, then sing the Benedicite and Pater Noster and a litany and stick thy knife into the wort, make it stick fast and go away; go again when day and night just divide; at the same period go first to church and cross thyself and commend thyself to God; then go in silence and, though anything soever of an awful sort or man meet thee, say not thou to him any word ere thou come to the wort which on the evening before thou markedst; then sing the Benedicite and the Pater Noster and a litany, delve up the wort, let the knife stick in it; go again as quickly as thou art able to church and let it lie under the altar with the knife; let it lie till the sun be up, wash it afterwards, and make into a drink with bishopwort and lichen off a crucifix; boil in milk thrice, thrice pour holy water upon it and sing over it the Pater Noster, the Credo and the Gloria in Excelsis Deo, and sing upon it a litany and score with a sword round about it on three sides a cross, and then after that let the man drink the wort; Soon it will be well with him."—*Leech Book*, III. 62.

The instructions for a horse or cattle that are elf-shot runs thus :—

[1] *Herb. Ap.*, I.

"If a horse or other neat be elf-shot take sorrel-seed or Scotch wax, let a man sing twelve Masses over it and put holy water on the horse or on whatsoever neat it be; have the worts always with thee. For the same take the eye of a broken needle, give the horse a prick with it, no harm shall come."—*Leech Book of Bald,* I. 88.

Another prescription for an elf-shot horse runs thus:—

"If a horse be elf-shot, then take the knife of which the haft is the horn of a fallow ox and on which are three brass nails, then write upon the horse's forehead Christ's mark and on each of the limbs which thou mayst feel at: then take the left ear, prick a hole in it in silence, this thou shalt do; then strike the horse on the back, then will it be whole.—And write upon the handle of the knife these words—

"Benedicite omnia opera Domini dominum.

"Be the elf what it may, this is mighty for him to amend."—*Leech Book of Bald,* I. 65.[1]

Closely allied to the doctrine of the elf-shot is that of "flying venom." It is, of course, possible to regard the phrase as the graphic Anglo-Saxon way of describing infectious diseases; but the various synonymous phrases, "the on-flying things," "the loathed things that rove through the land," suggest something of more malignant activity. As a recent leading article in *The Times* shows, we are as a matter of fact not much wiser than our Saxon ancestors as to the origin of an epidemic such as influenza.[2] Indeed, to talk of "catching" a cold or any infec-

[1] For "elf-shot" herbal remedies see also *Leech Book,* III. 1, 61, 64.

[2] "The visitation raises again questions which were so anxiously propounded three years ago. In what manner does an epidemic of this kind arise? How is it propagated? We are still to a great extent in the dark in regard to both these points. Indeed, it has recently been suggested that we do not 'catch' influenza at all, but that certain climatic or other conditions favour the multiplication on an important scale of micro-organisms normally present in the human air passages. It would be foolish to pretend

tious disease would have struck an Anglo-Saxon as ludicrous, mankind being rather the victims of "flying venom." In the alliterative lay in the *Lacnunga,* part of which is given below, the wind is described as blowing these venoms, which produced disease in the bodies on which they lighted, their evil effects being subsequently blown away by the magician's song and the efficacy of salt and water and herbs. This is generally supposed to be in its origin a heathen lay of great antiquity preserved down to Christian times, when allusions to the new religion were inserted. It is written in the Wessex dialect and is believed to be of the tenth century, but it is undoubtedly a reminiscence of some far older lay. The lay or charm is in praise of nine sacred herbs (one a tree)—mugwort, waybroad (plantain), stime (watercress), atterlothe (?), maythen (camomile), wergulu (nettle), crab apple, chervil and fennel.

> "These nine attack
> against nine venoms.
> A worm came creeping,
> he tore asunder a man.
> Then took Woden
> nine magic twigs,
> [&] then smote the serpent
> that he in nine [bits] dispersed.
> Now these nine herbs have power
> against nine magic outcasts
> against nine venoms
> & against nine flying things
> [& have might] against the loathed things
> that over land rove.
> Against the red venoms
> against the runlan [?] venom
> against the white venom
> against the blue [?] venom
> against the yellow venom
> against the green venom
> against the dusky venom
> against the brown venom
> against the purple venom.

---

to any opinion on a subject which is at present almost entirely speculative: yet the theory we have quoted may serve to show how complicated and difficult are the issues involved."—*The Times,* January 13, 1922.

Against worm blast
against water blast
against thorn blast
against thistle blast
Against ice blast
Against venom blast

. . . . . .

if any venom come
flying from east
or any come from north
[or any from south]
or any from west
over mankind
I alone know a running river
and the nine serpents behold [it]
All weeds must
now to herbs give way,
Seas dissolve
[and] all salt water
when I this venom
from thee blow." [1]

In the chapter in the *Leech Book of Bald* [2] containing the prescriptions sent by the Patriarch of Jerusalem to King Alfred, we find among the virtues of the "white stone" that it is "powerful against flying venom and against all uncouth things," and in another passage [3] that these venoms are particularly dangerous "fifteen nights ere Lammas and after it for five and thirty nights: leeches who were wisest have taught that in that month no man should anywhere weaken his body except there were a necessity for it." In the most ancient source of Anglo-Saxon medicine—the *Lacnunga*—we find the following "salve" for flying venom:—

"A salve for flying venom. Take a handful of hammer wort and a handful of maythe (camomile) and a handful of waybroad (plantain) and roots of water dock, seek those which

[1] Translation from Dr. Charles Singer's *Early English Magic and Medicine*. Proceedings of the British Academy.

[2] *Leech Book of Bald*, Book II. 64

[3] *Id*. Book I. 72. For other references to flying venom see *Leech Book of Bald*, I. 113; II. 65.

will float, and one eggshell full of clean honey, then take clean butter, let him who will help to work up the salve melt it thrice: let one sing a mass over the worts, before they are put together and the salve is wrought up." [1]

But it is in the doctrine of the worm as the ultimate source of disease that we are carried back to the most ancient of sagas. The dragon and the worm, the supreme enemy of man, which play so dominating a part in Saxon literature, are here set down as the source of all ill. In the alliterative lay in the *Lacnunga* the opening lines describe the war between Woden and the Serpent. Disease arose from the nine fragments into which he smote the serpent, and these diseases, blown by the wind, are counteracted by the nine magic twigs and salt water and herbs with which the disease is again blown away from the victim by the power of the magician's song. This is the atmosphere of the great earth-worm Fafnir in the Volsunga Saga and the dragon in all folk tales, the great beast with whom the heroes of all nations have contended. Further, it is noteworthy that not only in Anglo-Saxon medicine, but for many centuries afterwards, even minor ailments were ascribed to the presence of a worm—notably toothache. In the *Leech Book* we find toothache ascribed to a worm in the tooth (see *Leech Book*, II. 121). It is impossible in a book of this size to deal with the comparative folk lore of this subject, but in passing it is interesting to recall an incantation for toothache from the Babylonian cuneiform texts [2] in which we find perhaps the oldest example of this belief.

> " The Marshes created the Worm,
> Came the Worm and wept before Shamash,
> What wilt thou give me for my food?
> What wilt thou give me to devour?
>
> . . . . . .

[1] *Lacnunga*, 6.
[2] *Cuneiform Texts*, Part XVII. pl. 50.

Let me drink among the teeth
And set me on the gums,
That I may devour the blood of the teeth
And of the gums destroy their strength.
Then shall I hold the bolt of the door.

. . . . . .

So must thou say this, O Worm,
May Ea smite thee with the might of his fist."

Closely interwoven with these elements of Indo-Germanic origin we find the ancient Eastern doctrine which ascribes disease to demoniac possession. The exorcisms were originally heathen charms, and even in the *Leech Book* there are many interesting survivals of these, although Christian rites have to a large extent been substituted for them. Both mandrake and periwinkle were supposed to be endowed with mysterious powers against demoniacal possession. At the end of the description of the mandrake in the *Herbarium of Apuleius* there is this prescription :—

"For witlessness, that is devil sickness or demoniacal possession, take from the body of this same wort mandrake by the weight of three pennies, administer to drink in warm water as he may find most convenient—soon he will be healed."—*Herb. Ap.*, 32.

Of periwinkle we read :—

"This wort is of good advantage for many purposes, that is to say first against devil sickness and demoniacal possessions and against snakes and wild beasts and against poisons and for various wishes and for envy and for terror and that thou mayst have grace, and if thou hast the wort with thee thou shalt be prosperous and ever acceptable. This wort thou shalt pluck thus, saying, 'I pray thee, vinca pervinca, thee that art to be had for thy many useful qualities, that thou come to me glad blossoming with thy mainfulness, that thou outfit me so that I be shielded and ever prosperous and undamaged by poisons and by water;' when thou shalt pluck this wort thou shalt be clean of every uncleanness, and thou shalt pick it when the moon is nine nights old and eleven nights and thirteen nights and thirty nights and when it is one night old."—*Herb. Ap.*

In the treatment of disease we find that the material remedies, by which I mean remedies devoid of any mystic meaning, are with few exceptions entirely herbal. The herb drinks were made up with ale, milk or vinegar, many of the potions were made of herbs mixed with honey, and ointments were made of herbs worked up with butter. The most scientific prescription is that for a vapour bath,[1] and there are suggestions for what may become fashionable once more—herb baths. The majority of the prescriptions are fc common ailments, and one cannot help being struck by the number there are for broken heads, bleeding noses and bites of mad dogs. However ignorant one may be of medicine, it is impossible to read these old prescriptions without realising that our ancestors were an uncommonly hardy race, for the majority of the remedies would kill any of us modern

[1] The directions for the vapour bath are given in such a brief and yet forceful way that I cannot imagine anyone reading it without feeling at the end as though he had run breathlessly to collect the herbs, and then prepared the bath and finally made the ley of alder ashes to wash the unfortunate patient's head. Like all these cheerful Saxon prescriptions, this one ends with the comforting assurance "it will soon be well with him," and one wonders whether in this, as in many other cases, the patient got well in order to avoid his friends' ministrations. The prescription for a vapour bath made with herbs runs thus:—

"Take bramble rind and elm rind, ash rind, sloethorn, rind of apple tree and ivy, all these from the nether part of the trees, and cucumber, smear wort, everfern, helenium, enchanters nightshade, betony, marrubium, radish, agrimony. Scrape the worts into a kettle and boil strongly. When it hath strongly boiled remove it off the fire and seat the man over it and wrap the man up that the vapour may get up ı owhere, except only that the man may breathe; beathe him with these fomentations as long as he can bear it. Then have another bath ready for him, take an emmet bed all at once, a bed of those male emmets which at whiles fly, they are red ones, boil them in water, beathe him with it immoderately hot. Then make him a salve. Take worts of each kind of those above mentioned, boil them in butter, smear the sore limbs, they will soon quicken. Make him a ley of alder ashes, wash his head with this cold, it will soon be well with him, and let the man get bled every month when the moon is five and fifteen and twenty nights old."

MANDRAKE FROM A SAXON HERBAL

(Sloane 1975, folio 49*a*)

weaklings, even if in robust health when they were administered. At times one cannot help wondering whether in those days, as not infrequently happens now, the bulletin was issued that "the operation was quite successful, but the patient died of shock!" And, as further evidence of the old truth that there is nothing new under the sun, it is pleasant to find that doctors, even in Saxon days, prescribed "carriage exercise," and moreover endeavoured to sweeten it by allowing the patient to "lap up honey" first. This prescription runs thus :—

"Against want of appetite. Let them, after the night's fast, lap up honey, and let them seek for themselves fatigue in riding on horseback or in a wain or such conveyance as they may endure."—*Leech Book*, II. 7.

In the later herbals, "beauty" recipes are, as is well known, a conspicuous feature, but they find a place also in these old manuscripts. In the third book (the oldest part) of the *Leech Book* there is a prescription for sunburn which runs thus :—

"For sunburn boil in butter tender ivy twigs, smear therewith."—*Leech Book*, III. 29.

And in *Leech Book* II. we find this prescription :—

"That all the body may be of a clean and glad and bright hue, take oil and dregs of old wine equally much, put them into a mortar, mingle well together and smear the body with this in the sun."—*Leech Book*, II. 65.

Prescriptions for hair falling off are fairly numerous, and there are even two—somewhat drastic—prescriptions for hair which is too thick. Sowbread and watercress were both used to make hair grow, and in *Leech Book* I. there is this prescription :—

"If a man's hair fall off, work him a salve. Take the mickle wolf's bane and viper's bugloss and the netherward part of burdock, work the salve out of that wort and out of all these and out of that butter of which no water hath come. If hair fall off, boil the polypody fern and foment the head with that so warm. In case that a man be bald, Plinius the mickle leech saith this leechdom: 'Take dead bees, burn them to ashes, add oil upon that, seethe very long over gledes, then strain, wring out and take leaves of willow, pound them, pour the juice into the oil; boil again for a while on gledes, strain them, smear therewith after the bath."—*Leech Book*, I. 87.

The two prescriptions for hair which is too thick are in the same chapter :—

"In order that the hair may not wax, take emmets' eggs, rub them up, smudge on the place, never will any hair come up there." Again: "if hair be too thick, take a swallow, burn it to ashes under a tile and have the ashes shed on."

There are more provisions against diseases of the eye than against any other complaint, and it is probably because of the prevalence of these in olden days that we still have so many of the superstitions connected with springs of water. Both maythen (camomile) and wild lettuce were used for the eyes. In the following for mistiness of eyes there is a touch of pathos :—

"For mistiness of eyes, many men, lest their eyes should suffer the disease, look into cold water and then are able to see far. . . . The eyes of an old man are not sharp of sight, then shall he wake up his eyes with rubbings, with walkings, with ridings, either so that a man bear him or convey him in a wain. And they shall use little and careful meats and comb their heads and drink wormwood before they take food. Then shall a salve be wrought for unsharpsighted eyes; take pepper and beat it and a somewhat of salt and wine; that will be a good salve."

One prescription is unique, for the "herb" which one is directed to use is not to be found in any other herbal in existence. This is "rind from Paradise." There is a grim humour about the scribe's comment, and one cannot help wondering what was the origin of the prescription :—

"Some teach us against bite of adder, to speak one word 'faul.' It may not hurt him. Against bite of snake if the man procures and eateth rind which cometh out of Paradise, no venom will hurt him. Then said he that wrote this book that the rind was hard gotten."

These manuscripts are so full of word pictures of the treatment of disease that one feels if one were transported back to those days it would in most cases be possible to tell at a glance the "cures" various people were undergoing. Let us visit a Saxon hamlet and go and see the sick folk in the cottages. On our way we meet a man with a fawn's skin decorated with little bunches of herbs dangling from his shoulders, and we know that he is a sufferer from nightmare.[1] Another has a wreath of clove-wort tied with a red thread round his neck. He is a lunatic, but, as the moon is on the wane, his family hope that the wearing of these herbs will prove beneficial. We enter a dark one-roomed hut, the dwelling of one of the swineherds, but he is not at his work; for it seemed to him that his head turned about and that he was faring with turned brains. He had consulted the leech and, suggestion cures being then rather more common than now, the leech had advised him to sit calmly by his fireside with a linen cloth wrung out in spring water on his head and to wait till it was dry. He does so, and, to quote the words with which nearly all Saxon prescriptions end, we feel "it will soon be well with him." Let us wend our way to the cobbler, a sullen, taciturn man who finds his lively young wife's chatter unendurable. We find him looking more gloomy than usual, for he has eaten nothing all day and now sits moodily consuming a raw radish. But there is purpose in this. Does not the ancient leechdom say that, if a radish be eaten raw after fasting all day, no woman's chatter the next day can annoy? In another cottage we find that a patient suffering from elf-shot is to be smoked with the fumes of herbs. A huge

[1] *Leech Book*, I. 60.

quern stone which has been in the fire on the hearth all day is dragged out, the prepared herbs—wallwort and mugwort—are scattered upon it and also underneath, then cold water is poured on and the patient is reeked with the steam "as hot as he can endure it."[1] Smoking sick folk, especially for demoniac possession, is a world-wide practice and of very ancient origin. There is no space here to attempt to touch on the comparative folk lore of this subject. Moreover, fumigating the sick with herbs is closely akin to the burning of incense. Even in ancient Babylonian days fumigating with herbs was practised.[2] It was very common all through the Middle Ages in most parts of Europe, and that it has not even yet died out is shown by the extract from *The Times* given below.[3] I have purposely put in juxtaposition the translation of the ancient Babylonian tablet and the extract from *The Times*.

[1] *Lacnunga*, 48.

[2] In an incantation against fever we find the instruction :—

> "The sick man . . . thou shalt place
> . . . . . . thou shalt cover his face
> Burn cypress and herbs . . . . . .
> That the great gods may remove the evil
> That the evil spirit may stand aside
> . . . .
> May a kindly spirit a kindly genius be present."

R. Campbell Thompson, *Devils and Evil Spirits of Babylonia*, p. 29. See also p. 43. Cf. also Tobit vi. 7.

[3] *A Pomeranian Rite.*—An attempt was made a few days ago to cast a devil out of a woman living in a village of the Lauenberg district of Pomerania, on the Polish frontier. She appears to have been of a sour and somewhat hysterical temperament, and three of the village gossips came to the conclusion that she was a victim of diabolical possession and resolved to effect a cure by means of enchantment. They first of all gathered the herbs needed for the purpose in the forest at the proper conjunction of the stars. Then a tripod was formed of three chairs, and to these the patient was bound. Beneath her was fixed a pail of red-hot coal on which the herbs were scattered. As the fumes of the burning weeds veiled the victim the three neighbours crooned the prescribed exorcism. The louder the woman shrieked the louder they sang, and after the process had been continued long enough to prove effective, in their opinion, they ran away, believing that the devil would run out of the woman after them. She, however, continued to shriek. Her cries were heard by a man, who released her.—*The Times*, December 5, 1921.

It is noteworthy that not only human beings, but cattle and swine were smoked with the fumes of herbs. In the *Lacnunga*, for sick cattle we find—" Take the wort, put it upon gledes and fennel and hassuck and ' cotton ' and incense. Burn all together on the side on which the wind is. Make it reek upon the cattle. Make five crosses of hassuck grass, set them on four sides of the cattle and one in the middle. Sing about the cattle the Benedicite and some litanies and the Pater Noster. Sprinkle holy water upon them, burn about them incense and cotton and let someone set a value on the cattle, let the owner give the tenth penny in the Church for God, after that leave them to amend; do this thrice."—*Lacnunga*, 79.

" To preserve swine from sudden death sing over them four masses, drive the swine to the fold, hang the worts upon the four sides and upon the door, also burn them, adding incense and make the reek stream over the swine."—*Lacnunga*, 82.

Herbs used as amulets have always played a conspicuous part in folk medicine, and our Saxon ancestors used them, as all ancient races have used them, not merely to cure definite diseases but also as protection against the unseen powers of evil,[1] to preserve the eyesight, to cure lunacy, against weariness

[1] It is interesting to find the same beliefs amongst the ancient Babylonians.

" Fleabane on the lintel of the door I have hung
S. John's wort, caper and wheatears
With a halter as a roving ass
Thy body I restrain.
O evil spirit get thee hence
Depart O evil Demon.

. . . .

In the precincts of the house stand not nor circle round
' In the house will I stand,' say thou not,
' In the neighbourhood will I stand,' say thou not.
O evil spirit get thee forth to distant places
O evil Demon hie thee unto the ruins
Where thou standest is forbidden ground
A ruined desolate house is thy home
Be thou removed from before me, By Heaven be thou exorcised
By Earth be thou exorcised."

Trans. of Utukke Limnûte Tablet " B." R. C. Thompson, *Devils and Evil Spirits of Babylonia.*

when going on a journey, against being barked at by dogs, for safety from robbers, and in one prescription even to restore a woman stricken with speechlessness. The use of herbs as amulets to cure diseases has almost died out in this country, but the use of them as charms to ensure good luck survives to this day—notably in the case of white heather and four-leaved clover.

There is occasionally the instruction to bind on the herb with red wool. For instance, a prescription against headache in the third book of the *Leech Book* enjoins binding waybroad, which has been dug up without iron before sunrise, round the head "with a red fillet." Binding on with red wool is a very ancient and widespread custom.[1] Red was the colour sacred to Thor and it was also the colour abhorred not only by witches in particular but by all the powers of darkness and evil. An ancient Assyrian eye charm prescribes binding "pure strands of red wool which have been brought by the pure hand of . . . on the right hand," and down to quite recent times even in these islands tying on with red wool was a common custom.

Besides their use as amulets, we also find instructions for hanging herbs up over doors, etc., for the benefit not only of human beings but of cattle also. Of mugwort we read in the *Herbarium of Apuleius*, "And if a root of this wort be hung over the door of any house then may not any man damage the house."

"Of Croton oil plant. For hail and rough weather to turn them away. If thou hast in thy possession this wort which is named 'ricinus' and which is not a native of England, if thou hangest some seed of it in thine house or have it or its seed in any place whatsoever, it turneth away the tempestuousness of hail, and if thou hangest its seed on a ship, to that degree

[1] Sonny (*Arch. f. Rel.*, 1906, p. 525), in his article "Rote Farbe im Totenkulte," considers the use of red to be in imitation of blood. The instruction to bind on with red is found even in the *Grete Herball* of 1526. "Apium is good for lunatyke Folke yf it be bounde to the pacyentes heed with a lynen clothe dyed reed," etc.

wonderful it is, that it smootheth every tempest. This wort thou shalt take saying thus, 'Wort ricinus I pray that thou be at my songs and that thou turn away hails and lightning bolts and all tempests through the name of Almighty God who hight thee to be produced'; and thou shalt be clean when thou pluckest this herb."—*Herb. Ap.*, 176.

"Against temptation of the fiend, a wort hight red niolin, red stalk, it waxeth by running water; if thou hast it on thee and under thy head and bolster and over thy house door the devil may not scathe thee within nor without."—*Leech Book*, III. 58.

"To preserve swine from sudden death take the worts lupin, bishopwort, hassuck grass, tufty thorn, vipers bugloss, drive the swine to the fold, hang the worts upon the four sides and upon the door."—*Lacnunga*, 82.

The herbs in commonest use as amulets were betony, vervain, peony, yarrow, mugwort and waybroad (plantain). With the exception of vervain, no herb was more highly prized than betony. The treatise on it in the *Herbarium of Apuleius* is supposed to be an abridged copy of a treatise on the virtues of this plant written by Antonius Musa, physician to the Emperor Augustus. No fewer than twenty-nine uses of it are given, and in the Saxon translation this herb is described as being "good whether for a man's soul or his body." Vervain was one of the herbs held most sacred by the Druids and, as the herbals of Gerard and Parkinson testify, it was in high repute even as late as the sixteenth and seventeenth centuries. It has never been satisfactorily identified, though many authorities incline to the belief that it was verbena. In Druidical times libations of honey had to be offered to the earth from which it was dug, mystic ceremonies attended the digging of it and the plant was lifted out with the left hand. This uprooting had always to be performed at the rising of the dog star and when neither the sun nor the moon was shining. Why the humble waybroad should occupy so prominent a place in Saxon herb lore it is difficult to understand. It is one of the nine sacred herbs in the alliterative lay in the *Lacnunga*, and the epithets "mother of worts" and "open from eastwards" are applied to it. The latter curious epithet is also applied to it in *Lacnunga* 46,—"which spreadeth open towards the East." Waybroad has certainly wonderfully curative powers, especially for bee-stings, but otherwise it has long since fallen from its high estate. Peony throughout the Middle Ages was held in high repute for its protective powers, and even during the closing years of the last century country folk hung beads made of its roots round children's necks.[1] Yarrow is one of the aboriginal English plants, and from time immemorial it has been used in incantations and by witches. Country folk still regard it as one of our most valuable herbs, especially for rheumatism. Mugwort, which was held in repute throughout the Middle Ages for its efficacy against unseen powers of evil, is one of the nine sacred herbs in the alliterative lay in the *Lacnunga*, where it is described thus :—

"Eldest of worts
Thou hast might for three
And against thirty
For venom availest
For flying vile things,
Mighty against loathed ones
That through the land rove."

Harleian MS. 585.

With the notable exception of vervain, it is curious how little prominence is given in Saxon plant lore to the herbs which were held most sacred by the Druids, and yet it is scarcely credible that some of their wonderful lore should not have been assimilated. But in these manuscripts little or no importance attaches to mistletoe, holly, birch or ivy. There is no mention of mistletoe as a sacred herb.[2] We find some mention of selago,

[1] See W. G. Black, *Folk Medicine*.

[2] Even modern science has not yet succeeded in solving some of the mysteries connected with this remarkable plant. For instance, although the

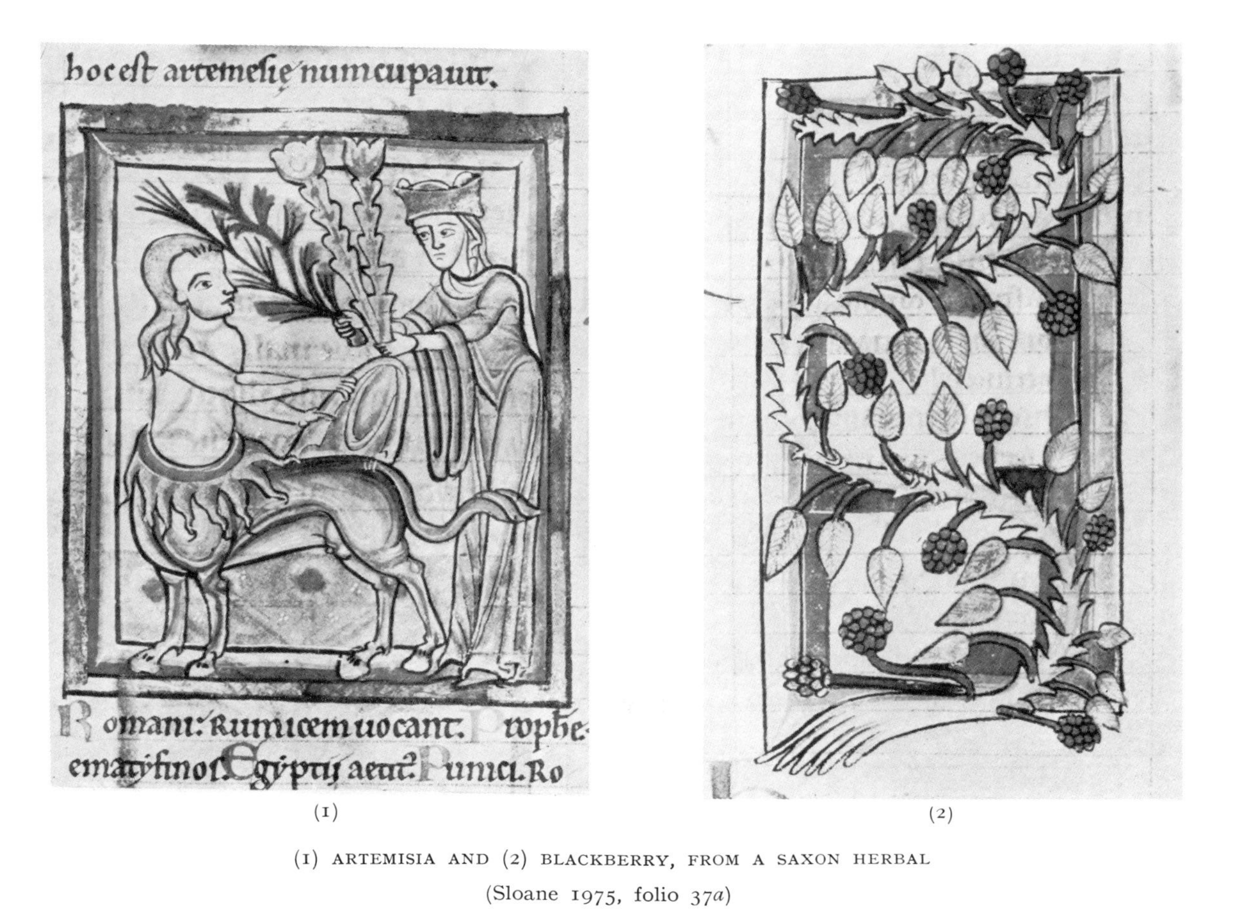

(1) (2)

(1) ARTEMISIA AND (2) BLACKBERRY, FROM A SAXON HERBAL

(Sloane 1975, folio 37*a*)

generally identified with *lycopodium selago*, of which Pliny tells us vaguely that it was "like savin." The gathering of it had to be accompanied in Druid days with mystic ceremonies. The Druid had his feet bare and was clad in white, and the plant could not be cut with iron, nor touched with the naked hand. So great were its powers that it was called "the gift of God." Nor is there any mention in Saxon plant lore of the use of *sorbus aucuparia*, which the Druids planted near their monolithic circles as protection against unseen powers of darkness. There is, however, one prescription which may date back to the Roman occupation of Britain. It runs thus: "Take nettles, and seethe them in oil, smear and rub all thy body therewith; the cold will depart away."[1] It has always been believed that one of the varieties of nettle (*Urtica pilulifera*) was introduced into England by the Roman soldiers, who brought the seed of it with them. According to the tradition, they were told that the cold in England was unendurable; so they brought these seeds in order to have a plentiful supply of nettles wherewith to rub their bodies and thereby keep themselves warm. Possibly this prescription dates back to that time.

From what hoary antiquity the charms and incantations which we find in these manuscripts have come down to us we cannot say. Their atmosphere is that of palæolithic cave-drawings, for they are redolent of the craft of sorcerers and they suggest those strange cave markings which no one can decipher. Who can say what lost languages are embedded in these unintelligible words and single letters, or what is their meaning? To what ancient ceremonies do they pertain, and who were the

---

apple and the pear are closely related, mistletoe very rarely grows on the pear tree, and there is no case on record of mistletoe planted on a pear tree by human hands surviving the stage of germination. There are, it is true, two famous mistletoe pears in this country—one in the garden of Belvoir Castle and the other in the garden of Fern Lodge, Malvern, but in both cases the seed was sown naturally. It grows very rarely on the oak, and this possibly accounts for the special reverence accorded by the Druids to the mistletoe oak.

[1] *Leech Book*, I. 81.

initiated who alone understood them? At present it is all mysterious, though perhaps one day we shall discover both their sources and their meaning. They show no definite traces of the Scandinavian rune-lays concerning herbs, though one of the charms is in runic characters. It is noteworthy that in the third book, which is evidently much older than the first two parts of the *Leech Book,* the proportion of heathen charms is exceptionally large. In one prescription we find the names of two heathen idols, Tiecon and Leleloth, combined with a later Christian interpolation of the names of the four gospellers. The charm is in runic characters and is to be followed by a prayer. Many of the mystic sentences are wholly incomprehensible, in others we find heathen names such as Lilumenne, in others a string of words which may be a corrupt form of some very ancient language. Thus a lay to be sung in case a man or beast drinks an insect runs thus:—" Gonomil, orgomil, marbumil, marbsai, tofeth," etc.[1]

If some of the charms have a malignant sound, others were probably as soothing in those days as those gems are still which have survived in our inimitable nursery rhymes.

For instance, the following has for us no meaning, but even in the translation it has something of the curious effect of the words in the original. A woman who cannot rear her child is instructed to say—" Everywhere I carried for me the famous kindred doughty one with this famous meat doughty one, so I will have it for me and go home."

In the *Lacnunga* there is a counting-out charm which is a mixture of an ancient heathen charm combined with a Christian rite at the end.

" Nine were Noddes sisters, then the nine came to be eight, and the eight seven, and the seven six, and the six five, and the five four, and the four three, and the three two, and the two one, and the one none. This may be medicine for thee from scrofula

[1] *Lacnunga,* 9.

and from worm and from every mischief. Sing also the Benedicite nine times."—*Lacnunga*, 95.[1]

One of the most remarkable narrative charms is that for warts copied below from the *Lacnunga*. It is to be sung first into the left ear, then into the right ear, then above the man's poll, then "let one who is a maiden go to him and hang it upon his neck, do so for three days, it will soon be well with him."

"Here came entering
A spider wight.
He had his hands upon his hams.
He quoth that thou his hackney wert.
Lay thee against his neck.
They began to sail off the land.
As soon as they off the land came, then began they to cool.
Then came in a wild beast's sister.
Then she ended
And oaths she swore that never could this harm the sick, nor him who could get at this charm, nor him who had skill to sing this charm. Amen. Fiat."—*Lacnunga*, 56.

Of the world-wide custom of charming disease from the patient and transferring it to some inanimate object we find numerous examples. This custom is not only of very ancient origin, but persisted until recent times even in this country. As commonly practised in out-of-the-way parts of Great Britain it was believed that the disease transferred to an inanimate object would be contracted by the next person who picked it up, but in the Saxon herbals we find an apparently older custom of transferring the disease to "running water" (suggestive of the Israelitish scapegoat), and also that of throwing the blood from the wound across the wagon way. These charms for transferring disease seem originally to have been associated

[1] This closely resembles a Cornish charm for a tetter.

"Tetter, tetter, thou hast nine brothers,
God bless the flesh and preserve the bone;
Perish thou, tetter, and be thou gone.
Tetter, tetter, thou hast eight brothers."

Thus the verses are continued until tetter having "no brother" is ordered to be gone.—R. Hunt, *Popular Romances of the West of England*, p. 414.

with a considerable amount of ceremonial. For instance, in those to cure the bite of a hunting spider we find that a certain number of scarifications are to be struck (and in both cases an odd number—three and five); in the case of the five scarifications, "one on the bite and four round about it," the blood is to be caught in "a green spoon of hazel-wood," and the blood is to be thrown "in silence" over a wagon way. In the *Lacnunga* there are traces of the actual ceremonial of transferring the disease, and the Christian prayer has obviously been substituted for an older heathen one. The charm is in unintelligible words and is followed by the instruction, "Sing this nine times and the Pater Noster nine times over a barley loaf and give it to the horse to eat." In a "salve against the elfin race" it is noticeable that the herbs, after elaborate preparation, are not administered to the patient at all, but are thrown into running water.

"A salve against the elfin race and nocturnal goblin visitors: take wormwood, lupin. . . . Put these worts into a vessel, set them under the altar, sing over them nine masses, boil them in butter and sheep's grease, add much holy salt, strain through a cloth, throw the worts into running water."—*Leech Book,* III. 61.

One charm in the *Lacnunga* which is perhaps not too long to quote speaks of some long-lost tale. It appears to be a fragment of a popular lay, and one wonders how many countless generations of our ancestors sang it, and what it commemorates :—

"Loud were they loud, as over the land they rode,
Fierce of heart were they, as over the hill they rode.
Shield thee now thyself; from this spite thou mayst escape thee!
Out little spear if herein thou be!
Underneath the linden stood he, underneath the shining shield,
While the mighty women mustered up their strength;
And the spears they send screaming through the air!
Back again to them will I send another.
Arrow forth a-flying from the front against them;
Out little spear if herein thou be!
Sat the smith thereat, smoke a little seax out.
Out little spear if herein thou be!
Six the smiths that sat there— making slaughter-spears:

Out little spear, in be not spear!
If herein there hide flake of iron hard,
Of a witch the work, it shall melt away.
Wert thou shot into the skin, or shot into the flesh,
Wert thou shot into the blood, or shot into the bone,
Wert thou shot into the limb— never more thy life be teased!
If it were the shot of Esa, or it were of elves the shot
Or it were of hags the shot; help I bring to thee.
This to boot for Esa-shot, this to boot for elfin-shot,
This to boot for shot of hags! Help I bring to thee.
Flee witch to the wild hill top . . . . . .
But thou—be thou hale, and help thee the Lord."

Who were these six smiths and who were the witches? One thinks of that mighty Smith Weyland in the palace of Nidad king of the Niars, of the queen's fear of his flashing eyes and the maiming of him by her cruel orders, and of the cups he made from the skulls of her sons and gems from their eyes. We think of these as old tales, but instinct tells us that they are horribly real. We may not know how that semi-divine smith made himself wings, but that he flew over the palace and never returned we do not doubt for an instant. To the fairy stories which embody such myths children of unnumbered generations have listened, and they demand them over and over again because they, too, are sure that they are real.

Nor is the mystery of numbers lacking in these herbal prescriptions, particularly the numbers three and nine. In the alliterative lay of the nine healing herbs this is very conspicuous. Woden, we are told, smote the serpent with nine magic twigs, the serpent was broken into nine parts, from which the wind blew the nine flying venoms. There are numerous instances of the patient being directed to take nine of each of the ingredients or to take the herb potion itself for three or nine days. Or it is directed that an incantation is to be said or sung three or nine times, or that three or nine masses are to be sung over the herbs. This mystic use of three and nine is conspicuous in the following prescription:—

" Against dysentery, a bramble of which both ends are in

the earth take the newer root, delve it up, cut up nine chips with the left hand and sing three times the Miserere mei Deus and nine times the Pater Noster, then take mugwort and everlasting, boil these three worts and the chips in milk till they get red, then let the man sip at night fasting a pound dish full . . . let him rest himself soft and wrap himself up warm; if more need be let him do so again, if thou still need do it a third time, thou wilt not need oftener."—*Leech Book*, II. 65.

The leechdom for the use of dwarf elder against a snake-bite runs thus :—[1]

"For rent by snake take this wort and ere thou carve it off hold it in thine hand and say thrice nine times Omnes malas bestias canto, that is in our language Enchant and overcome all evil wild deer; then carve it off with a very sharp knife into three parts."—*Herb. Ap.*, 93.

Some of the most remarkable passages in the manuscripts are those concerning the ceremonies to be observed both in the picking and in the administering of herbs. What the mystery of plant life which has so deeply affected the minds of men in all ages and of all civilisations meant to our ancestors, we can but dimly apprehend as we study these ceremonies. They carry us back to that worship of earth and the forces of Nature which prevailed when Woden was yet unborn. That Woden was the chief god of the tribes on the mainland is indisputable, but even in the hierarchy of ancestors reverenced as semi-divine the Saxons themselves looked to Sceaf rather than to Woden, who himself was descended from Sceaf. There are few more haunting legends than that of our mystic forefather, the little boy asleep on a sheaf of corn who, in a richly adorned vessel which moved neither by sails nor oars, came to our people out of the great deep and was hailed by them as their king. Did not Alfred himself claim him as his primeval progenitor, the founder of

[1] For further instances of the mystic use of three and nine see also *Leech Book*, I. 45, 47, 67.

our race? There is no tangible link between his descendant Woden and the worship of earth, but the sheaf of corn, the symbol of Sceaf, carries us straight back to Nature worship. Sceaf takes his fitting place as the semi-divine ancestor with the lesser divinities such as Hrede and Eostra, goddess of the radiant dawn. It is to this age that the ceremonies in the picking of the herbs transport us, to the mystery of the virtues of herbs, the fertility of earth, the never-ceasing conflict between the beneficent forces of sun and summer and the evil powers of the long, dark northern winters. Closely intertwined with Nature worship we find the later Christian rites and ceremonies. For the new teaching did not oust the old, and for many centuries the mind of the average man halted half-way between the two faiths. If he accepted Christ he did not cease to fear the great hierarchy of unseen powers of Nature, the worship of which was bred in his very bone. The ancient festivals of Yule and Eostra continued under another guise and polytheism still held its sway. The devil became one with the gloomy and terrible in Nature, with the malignant elves and dwarfs. Even with the warfare between the beneficent powers of sun and the fertility of Nature and the malignant powers of winter, the devil became associated. Nor did men cease to believe in the Wyrd, that dark, ultimate fate goddess who, though obscure, lies at the back of all Saxon belief. It was in vain that the Church preached against superstitions. Egbert, Archbishop of York, in his Penitential, strictly forbade the gathering of herbs with incantations and enjoined the use of Christian rites, but it is probable that even when these manuscripts were written, the majority at least of the common folk in these islands, though nominally Christian, had not deserted their ancient ways of thought.[1]

[1] St. Eloy, in a sermon preached in A.D. 640, also forbade the enchanting of herbs:—

"Before all things I declare and testify to you that you shall observe none of the impious customs of the pagans, neither sorcerers, nor diviners, nor soothsayers, nor enchanters, nor must you presume for any cause to enquire of them. . . . Let none regulate the beginning of any piece of work

When the Saxon peasant went to gather his healing herbs he may have used Christian prayers [1] and ceremonies, but he did not forget the goddess of the dawn. It is noteworthy how frequently we find the injunction that the herbs must be picked at sunrise or when day and night divide, how often stress is laid upon looking towards the east, and turning " as the sun goeth from east to south and west." In many there is the instruction that the herb is to be gathered " without use of iron " or " with gold and with hart's horn " (emblems of the sun's rays). It is curious how little there is of moon lore. In some cases the herbs are to be gathered in silence, in others the man who gathers them is not to look behind him—a prohibition which occurs frequently in ancient superstitions. The ceremonies are all mysterious and suggestive, but behind them always lies the ancient ineradicable worship of Nature. To what dim past does that cry, " Erce, Erce, Erce, Mother of Earth " carry us?

" Erce, Erce, Erce, Mother of Earth!
May the All-Wielder, Ever Lord grant thee

---

by the day or by the moon. Let none trust in nor presume to invoke the names of dæmons, neither Neptune, nor Orcus, nor Diana, nor Minerva, nor Geniscus nor any other such follies. . . . Let no Christian place lights at the temples or the stones, or at fountains, or at trees, or at places where three ways meet. . . . Let none presume to hang amulets on the neck of man or beast. . . . Let no one presume to make lustrations, nor to enchant herbs, nor to make flocks pass through a hollow tree, or an aperture in the earth; for by so doing he seems to consecrate them to the devil. Let none on the kalends of January join in the wicked and ridiculous things, the dressing like old women or like stags, nor make feasts lasting all night, nor keep up the custom of gifts and intemperate drinking. Let no one on the festival of St. John or on any of the festivals join in the solstitia or dances or leaping or caraulas or diabolical songs."—From a sermon preached by St. Eloy in A.D. 640.

[1] A Christian prayer for a blessing on herbs runs thus:—

" Omnipotens sempiterne deus qui ab initio mundi omnia instituisti et creasti tam arborum generibus quam herbarum seminibus quibus etiam benedictione tua benedicendo sanxisti eadem nunc benedictione olera aliosque fructus sanctificare ac benedicere digneris ut sumentibus ex eis sanitatem conferant mentis et corporis ac tutelam defensionis eternamque uitam per saluatorem animarum dominum nostrum iesum christum qui uiuit et regnat dominus in secula seculorum. Amen."

Acres a-waxing, upwards a-growing
Pregnant [with corn] and plenteous in strength;
Hosts of [grain] shafts and of glittering plants!
Of broad barley the blossoms
And of white wheat ears waxing,
Of the whole earth the harvest!
Let be guarded the grain against all the ills
That are sown o'er the land by the sorcery men,
Nor let cunning women change it nor a crafty man."

And that other ancient verse :—

"Hail be thou, Earth, Mother of men!
In the lap of the God be thou a-growing!
Be filled with fodder for fare-need of men!"

It is of these two invocations that Stopford Brooke (whose translations I have used) writes: "These are very old heathen invocations used, I daresay, from century to century and from far prehistoric times by all the Teutonic farmers. Who 'Erce' is remains obscure. But the Mother of Earth seems to be here meant, and she is a person who greatly kindles our curiosity. To touch her is like touching empty space, so far away is she. At any rate some Godhead or other seems here set forth under her proper name. In the Northern Cosmogony, Night is the Mother of Earth. But Erce cannot be Night. She is (if Erce be a proper name) bound up with agriculture. Grimm suggests Eorce, connected with the Old High German 'erchan' = simplex. He also makes a bold guess that she may be the same as a divine dame in Low Saxon districts called Herke or Harke, who dispenses earthly goods in abundance, and acts in the same way as Berhta and Holda—an earth-goddess, the lady of the plougher and sower and reaper. In the Mark she is called Frau Harke. Montanus draws attention to the appearance of this charm in a convent at Corvei, in which this line begins—'Eostar, Eostar, eordhan modor.' . . . The name remains mysterious. The song breathes the pleasure and worship of ancient tillers of the soil in the labours of the earth and in the goods the mother gave. It has grown, it seems, out of the breast of earth herself; earth is here the Mother of Men. The surface of earth is the lap of the Goddess; in her womb let all growth be plentiful. Food is in her for the needs of men. 'Hail be thou, Earth!' I daresay this hymn was sung ten thousand years ago by the early Aryans on the Baltic coast."

Even in a twelfth-century herbal we find a prayer to Earth, and it is so beautiful that I close this chapter with it:—

"Earth,[1] divine goddess, Mother Nature who generatest all things and bringest forth anew the sun which thou hast given to the nations; Guardian of sky and sea and of all gods and powers and through thy power all nature falls silent and then sinks in sleep. And again thou bringest back the light and chasest away night and yet again thou coverest us most securely with thy shades. Thou dost contain chaos infinite, yea and winds and showers and storms; thou sendest them out when thou wilt and causest the seas to roar; thou chasest away the sun and arousest the storm. Again when thou wilt thou sendest forth the joyous day and givest the nourishment of life with thy eternal surety; and when the soul departs to thee we return. Thou indeed art duly called great Mother of the gods; thou conquerest by thy divine name. Thou art the source of the strength of nations and of gods, without thee nothing can be brought to perfection or be born; thou art great queen of the gods. Goddess! I adore thee as divine; I call upon thy name; be pleased to grant that which I ask thee, so shall I give thanks to thee, goddess, with one faith.

"Hear, I beseech thee, and be favourable to my prayer. Whatsoever herb thy power dost produce, give, I pray, with goodwill to all nations to save them and grant me this my medicine. Come to me with thy powers, and howsoever I may use them may they have good success and to whomsoever I may give them. Whatever thou dost grant it may prosper. To thee all things return. Those who rightly receive these herbs

[1] Translation from *Early English Magic and Medicine* by Dr. Charles Singer. Proceedings of the British Academy, Vol. IV.

FROM A SAXON HERBAL

(Harl. 1585, folio 19*a*)

from me, do thou make them whole. Goddess, I beseech thee; I pray thee as a suppliant that by thy majesty thou grant this to me.

"Now I make intercession to you all ye powers and herbs and to your majesty, ye whom Earth parent of all hath produced and given as a medicine of health to all nations and hath put majesty upon you, be, I pray you, the greatest help to the human race. This I pray and beseech from you, and be present here with your virtues, for she who created you hath herself promised that I may gather you into the goodwill of him on whom the art of medicine was bestowed, and grant for health's sake good medicine by grace of your powers. I pray grant me through your virtues that whatsoe'er is wrought by me through you may in all its powers have a good and speedy effect and good success and that I may always be permitted with the favour of your majesty to gather you into my hands and to glean your fruits. So shall I give thanks to you in the name of that majesty which ordained your birth."

CHAPTER II

LATER MANUSCRIPT HERBALS AND THE EARLY PRINTED HERBALS

"Spryngynge tyme is the time of gladnesse and of love; for in Sprynging time all thynge semeth gladde; for the erthe wexeth grene, trees burgynne [burgeon] and sprede, medowes bring forth flowers, heven shyneth, the see resteth and is quyete, foules synge and make theyr nestes, and al thynge that semed deed in wynter and widdered, ben renewed, in Spryngyng time."—BARTHOLOMÆUS ANGLICUS, *circ.* 1260.

BETWEEN the Anglo-Saxon herbals and the early printed herbals there is a great gulf. After the Norman Conquest the old Anglo-Saxon lore naturally fell into disrepute, although the Normans were inferior to the Saxons in their knowledge of herbs. The learned books of the conquerors were written exclusively in Latin, and it is sad to think of the number of beautiful Saxon books which must have been destroyed, for when the Saxons were turned out of their own monasteries the Normans who supplanted them probably regarded books written in a language they did not understand as mere rubbish. Much of the old Saxon herb lore is to be found in the leech books of the Middle Ages, but, with one notable exception, no important original treatise on herbs by an English writer has come down to us from that period. The vast majority of the herbal MSS. are merely transcriptions of Macer's herbal, a mediæval Latin poem on the virtues of seventy-seven plants, which is believed to have been written in the tenth century. The popularity of this poem is shown by the number of MSS. still extant. It was translated into English as early as the twelfth century with the addition of "A fewe herbes wyche Macer tretyth not." [1] In 1373 it was translated by John Lelamoure, a schoolmaster

[1] See Bibliography of English MS. Herbals.

of Hertford. On folio 55 of the MS. of this translation is the inscription, "God gracious of grauntis havythe yyeue and ygrauted vertuys in woodys stonys and herbes of the whiche erbis Macer the philosofure made a boke in Latyne the whiche boke Johannes Lelamoure scolemaistre of Herforde est, they he unworthy was in the yere of oure Lorde a. m. ccc. lxxiij tournyd in to Ynglis." Macer's herbal is also the basis of a treatise in rhyme of which there are several copies in England and one in the Royal Library at Stockholm. This treatise, which deals with twenty-four herbs, begins thus quaintly—

> "Of erbs xxiiij I woll you tell by and by
> Als I fond wryten in a boke at I in boroyng toke
> Of a gret ladys preste of gret name she barest."

The poem begins with a description of betony, powerful against "wykked sperytis," and then treats, amongst other herbs, of the virtues of centaury, marigold, celandine, pimpernel, motherwort, vervain, periwinkle, rose, lily, henbane, agrimony, sage, rue, fennel and violet. It is pleasant to find the belief that only to look on marigolds will draw evil humours out of the head and strengthen the eyesight.

> "Golde [marigold] is bitter in savour
> Fayr and ȝelw [yellow] is his flowur
> Ye golde flour is good to sene
> It makyth ye syth bryth and clene
> Wyscely to lokyn on his flowris
> Drawyth owt of ye heed wikked hirores [humours].
>
> . . . .
>
> Loke wyscely on golde erly at morwe [morning]
> Yat day fro feueres it schall ye borwe:
> Ye odour of ye golde is good to smelle."

The instructions for the picking of this joyous flower are given at length. It must be taken only when the moon is in the sign of the Virgin, and not when Jupiter is in the ascendant, for then the herb loses its virtue. And the gatherer, who must be out of deadly sin, must say three Pater Nosters and three Aves. Amongst its many virtues we find that it gives the wearer a vision of anyone who has robbed him. The virtues of vervain also are many; it must be picked "at Spring of day" in "ye monyth of May." Periwinkle is given its beautiful old name "joy of the ground" ("men calle it ye Juy of Grownde") and the description runs thus :—

> "Parwynke is an erbe grene of colour
> In tyme of May he beryth blo flour,
> His stalkys ain [are] so feynt [weak] and feye
> Yet never more growyth he heye [high]."

Under sage we find the old proverb—"How can a man die who has sage in his garden?"

> "Why of seknesse deyeth man
> Whill sawge [sage] in gardeyn he may han."

A manuscript of exceptional interest is one describing the virtues of rosemary which was sent by the Countess of Hainault to her daughter Philippa, Queen of England, and apart from its intrinsic interest it is important from the fact that it is obviously the original of the very poetical discourse on rosemary in the first printed English herbal, commonly known as Banckes's herbal. Moreover, in this MS. there is recorded an old tradition which I have not found in any other herbal, but which is still current amongst old-fashioned country folk, namely, that rosemary "passeth not commonly in highte the highte of Criste whill he was man on Erthe," and that when the plant attains the age of thirty-three years it will increase in breadth but not in height. It is the oldest MS. in which we find many other beliefs about rosemary that still survive in England. There is a tradition that Queen Philippa's mother sent the first plants of rosemary to England, and in a copy of this MS. in the library of Trinity College, Cambridge, the translator, "danyel bain," says that rosemary was unknown in England until the Countess of Hainault sent some to her daughter.

The only original treatise on herbs written by an Englishman during the Middle Ages was that by Bartholomæus Anglicus, and on the plant-lover there are probably few of the mediæval writers who exercise so potent a spell. Even in the thirteenth century, that age of great men, Bartholomew the Englishman ranked with thinkers such as Roger Bacon, Thomas Aquinas and Albertus Magnus. He was accounted one of the greatest theologians of his day, and if his lectures on theology were as simple as his writings on herbs, it is easy to understand why they were thronged and why his writings were so eagerly studied, not only in his lifetime but for nearly three centuries afterwards. A child could understand his book on herbs, for, being great, he was simple. But although his work *De Proprietatibus Rerum* (which contains nineteen books) was the source of common information on Natural History throughout the Middle Ages, and was one of the books hired out at a regulated price by the scholars of Paris, we know very little of the writer. He spent the greater part of his life in France and Saxony, but he was English born and was always known as Bartholomæus Anglicus.[1] We know that he studied in Paris and entered the French province of the Minorite Order, and later he became one of the most renowned professors of theology in Paris. In 1230 a letter was received from the general of the Friars Minor in the new province of Saxony asking the provincial of France to send Bartholomew and another Englishman to help in the work of that province, and the former subsequently went there. We do not know the exact date of *De Proprietatibus Rerum*, but it must have been written about

[1] He is sometimes erroneously called Bartholomew de Glanville. Leland, without citing any authority, called him de Glanville. Bale copied Leland in 1557 and added a list of writings wrongly attributed to Bartholomew. Quétif and Echard give detailed reasons in pointing out Leland's error. The Parmese chronicler, Salimbene, writing in 1283, refers to him as Bartholomæus Anglicus, and John de Trittenheim, Abbot of Sparheim (end of fifteenth century), speaks of him as "Bartholomeus natione Anglicus." M. Leopold Delisle endeavoured to claim him as a Frenchman, but although he spent the greater part of his life abroad, he was always distinguished as "Bartholomæus Anglicus." That he was a Minorite "de provincia Francia" is no proof that he was a Frenchman. Batman (1582), on the authority of Bale, describes Bartholomæus as being "of the noble familie of the Earles of Suffolk."

the middle of the thirteenth century; for, though it cites Albertus Magnus, who was teaching in Paris in 1248, there is no mention of any of the later authorities, such as Thomas Aquinas, Roger Bacon and Vincent de Beauvais. It was certainly known in England as early as 1296, for there is a copy of that date at Oxford, and there still exist both in France and in England a considerable number of other manuscript copies, most of which date from the latter part of the thirteenth century and the early part of the fourteenth. The book was translated into English in 1398 by John de Trevisa,[1] chaplain to Lord Berkeley and vicar of Berkeley, and Bartholomew could scarcely have been more fortunate in his translator. At the end of his translation, Trevisa writes thus :—

"Endlesse grace blysse thankyng and praysyng unto our Lorde God Omnipotent be gyuen, by whoos ayde and helpe this translacon was endyd at Berkeleye the syxte daye of Feuerer the yere of our Lorde MCCCLXXXXVIII the yere of yᵉ reyne of Kynge Rycharde the seconde after the Conqueste of Englonde XXII. The yere of my lordes aege, syre Thomas, Lorde of Berkeleye that made me to make this Translacōn XLVII."

Salimbene shows that the book was known in Italy in 1283, and there are two MS. copies in the Bibliothèque Nationale of Paris, of which the earliest is dated 1297. Before Trevisa made his English translation, it had been translated into French by Jehan Corbichon, in 1372, for Charles V. of France.

The book was first printed at Basle about 1470, and the esteem in which it was held may be judged from the fact that it went through at least fourteen editions before 1500, and besides the English and French translations it was also translated into Spanish and Dutch. The English translation was

[1] John de Trevisa, a Cornishman, was a Fellow of Exeter College, Oxford, and subsequently of Queen's College. He afterwards became chaplain to Lord Berkeley and vicar of Berkeley.

first printed by Caxton's famous apprentice, Wynken de Worde.[1] The translator in a naïve little introductory poem says that, just as he had looked as a child to God to help him in his games, so now he prays Him to help him in this book.

> " C[?]Rosse was made all of red .
> In the begynning of my boke .
> That is called, god me sped .
> In the fyrste lesson that j toke .
> Thenne I learned a and b .
> And other letters by her names .
> But alway God spede me .
> Thought me nedefull in all games .
> Yf I played in felde, other medes .
> Stylle other wyth noyse .
> I prayed help in all my dedes .
> Of him that deyed upon the croys .
> Now dyuerse playes in his name .
> I shall lette passe forth and far .
> And aventure to play so long game .
> Also I shall spare .
> Wodes, medes and feldes .
> Place that I have played inne .
> And in his name that all thīg weldes .
> This game j shall begynne. .
> And praye helpe conseyle and rede .
> To me that he wolde sende .
> And this game rule and lede .
> And brynge it to a good ende. ."

And in the preface Trevisa addresses his readers thus: " Merveyle not, ye witty and eloquent reders, that I thȳne of wytte and voyde of cunning have translatid this boke from latin to our vulgayre language as a thynge profitable to me and peradventure to many other, whych understonde not latyn nor have not the knowledge of the proprytees of thynges."

The seventeenth book of *De Proprietatibus Rerum* is on herbs and their uses, and it is full of allusions to the classical writers on herbs—Aristotle, Dioscorides and Galen—but the descriptions of the plants themselves are original and charming.

[1] Wynkyn de Worde's real name was Jan van Wynkyn (de Worde being merely a place-name), and in the sacrist's rolls of Westminster Abbey, 1491–1500, he figures as Johannes Wynkyn.

There is no record to show that Bartholomew the Englishman was a gardener, but we can hardly doubt that the man who described flowers with such loving care possessed a garden and worked in it. The *Herbarius zu Teutsch* might have been written in a study, but there is fresh air and the beauty of the living flowers in Bartholomew's writings. Of the lily he says: "The Lely is an herbe wyth a whyte floure. And though the levys of the floure be whyte yet wythen shyneth the lyknesse of golde." Bartholomew may have known nothing of the modern science of botany, but he knew how to describe not only the lily, but also the atmosphere of the lily, in a word-picture of inimitable simplicity and beauty. One feels instinctively that only a child or a great man could have written those lines. And is there not something unforgettable in these few words on the unfolding of a rose—"And whāne they [the petals] ben full growen they sprede theymselues ayenst the sonne rysynge"?

The chapter on the rose is longer than most, and is so delightful that I quote a considerable part of it. "The rose of gardens is planted and sette and tylthed as a vyne. And if it is forgendred and not shred and pared and not clensed of superfluyte: thēne it gooth out of kynde and chaungeth in to a wylde rose. And by oft chaunging and tylthing the wylde rose torneth and chaūgith into a very rose. And the rose of ye garden and the wylde rose ben dyuers in multitude of floures: smelle and colour: and also in vertue. For the leves of the wylde rose ben fewe and brode and whytyssh: meddlyd wyth lytyll rednesse: and smellyth not so wel as the tame rose, nother is so vertuous in medicyn. The tame rose hath many leuys sette nye togyder: and ben all red, other almost white: w^t wonder good smell. . . . And the more they ben brused and broken: the vertuoūser they ben and the better smellynge. And springeth out of a thorne that is harde and rough: netheles the Rose folowyth not the kynde of the thorne: But she arayeth her thorn wyth fayr colour and good smell. Whan ye rose begynneth to sprynge it is closed in a knoppe wyth grenes:

WOODCUT OF TREES AND HERBS FROM THE SEVENTEENTH BOOK OF "DE PROPRIETATIBUS RERUM"

Printed by Wynkyn de Worde (1495)

and that knoppe is grene. And whañe it swellyth thenne spryngeth out harde leuys and sharpe. . . . And whañe they ben full growen they sprede theymselues ayenst the sonne rysynge. And for they ben tendre and feble to holde togyder in the begynnynge; theyfore about those smale grene leuys ben nyghe the red and tendre leuys . . . and ben sette all aboute. And in the mydill thereof is seen the sede small and yellow wyth full gode smell."

There follows a description, too long to quote here, of the growth of the rose hip, which ends with the remark: "But they ben not ful good to ete for roughnesse that is hyd wythin. And greuyth [grieveth] wythin his throte that ete thereof." . . . "Among all floures of the worlde," he continues, "the floure of the rose is cheyf and beeryth ye pryse. And by cause of vertues and swete smelle and savour. For by fayrnesse they fede the syghte: and playseth the smelle by odour, the touche by softe handlynge. And wythstondeth and socouryth by vertue ayenst many syknesses and euylles." A delicious recipe is given for Rose honey. "Rose shreede smalle and sod in hony makyth that hony medycynable wyth gode smelle: And this comfortyeth and clenseth and defyeth gleymy humours."

Of the violet we read: "Violet is a lytyll herbe in substaunce and is better fresshe and newe than whan it is olde. And the floure thereof smellyth moost. . . . And the more vertuous the floure thereof is, ye more it bendyth the heed thereof doũwarde. Also floures of spryngynge tyme spryngeth fyrste and sheweth somer. The lytylnes thereof in substaunce is nobly rewarded in gretnesse of sauour and of vertue."

Bartholomew's descriptions of flowers are usually brief, and there is a clarity and vividness about them which give them a charm peculiarly their own. How fresh and English, for instance, is his chapter on the apple. I have never before seen the taste of an apple described as "merry," but how true the description is! "Malus the Appyll tree is a tree yt bereth apples and is a grete tree in itself . . . . it is more short than other trees of the wood wyth knottes and rinelyd Rynde. And makyth shadowe wythe thycke bowes and braunches: and fayr with dyuers blossomes, and floures of swetnesse and lykynge: with goode fruyte and noble. And is gracious in syght and in taste and vertuous in medecyne . . . some beryth sourysh fruyte and harde and some ryght soure and some ryght swete, with a good savoure and mery." The descriptions of celandine and broom are also characteristic. "Celidonia is an herbe w^t^ yelowe floures, the frute smorcheth them that it towchyth. And hyghte Celidonia for it spryngeth, other blomyth, in the comynge of swalowes. . . . It hyȝt celidonia for it helpith swallowes birdes yf their eyen be hurte other (or) blynde." "Genesta hath that name of bytterness for it is full of bytter to mannes taste. And is a shrubbe that growyth in a place that is forsaken, stony and untylthed. Presence thereof is wytnesse that the grounde is bareyne and drye that it groweth in. And hath many braunches knotty and hard. Grene in wynter and yelowe floures in somer thyche [the which] wrapped with heuy smell and bitter sauour. And ben netheles moost of vertue." Bartholomew gives the old mandrake legend in full, though he adds, "it is so feynd of churles others of wytches," and he also writes of its use as an anæsthetic.[1] Further, he records two other beliefs about the mandrake which I have never found in any other English herbal—namely, that while uprooting it one must beware of contrary winds, and that one must go on digging for it until sunset. "They that dygge mandragora be besy to beware of contrary wyndes whyle they digge. And maken circles abowte with a swerder and abyde with the dyggynge unto the sonne goynge downe."

But apart from herbs and their uses, the book *De herbis* is full of fleeting yet vigorous pictures of the homely everyday side of mediæval life. Bartholomew, being one of the greatest men of his century, writes of matters in which the simplest

[1] "The rind thereof medled with wine . . . gene to them to drink that shall be cut in their body for they should slepe and not fele the sore knitting."

of us are interested. He tells us of the feeding of swine with acorns. Of the making and baking of bread (including the thrifty custom of mixing cooked beans with the flour "to make the brede the more hevy"). Incidentally, and with all due respect, it may be remarked that he had no practical knowledge of this subject, his vivid description being obviously that of an interested spectator. There is an airy masculine vagueness about the conclusion of the whole matter of bread-making—"and at last after many travailes, man's lyfe is fedde and sustained therewith." He tells us of the use of laurel leaves to heal bee and wasp stings and to keep books and clothes from "moths and other worms," of the making of "fayre images" and of boxes wherein to keep "spycery" from the wood of the box-tree. Of the making of trestle tables "areared and set upon feet," of playing boards "that men playe on at the dyes [dice] and other gamys. And this maner of table is double and arrayd wyth dyerse colours." Of the making of writing tables, of wood used for flooring that "set in solar floors serue all men and bestys y[t] ben therein, and ben treden of alle men and beestys that come therein," and so strong that "they bende not nor croke [crack] whan they ben pressyd w[t] heuy thynges layd on them." And also of boards used for ships, bridges, hulks and coffers, and "in shypbreche [shipwreck] men fle to bordes and ben ofte sauyd in peryll." Of the building of houses with roofs of "trees stretchyd from the walles up to the toppe of ye house," with rafters "stronge and square and hewen playne," and of the covering of strawe and thetche [thatch]." Of the making of linen from the soaking of the flax in water till it is dried and turned in the sun and then bound in "praty bundels" and "afterward knockyd, beten and brayd and carflyd, rodded and gnodded; ribbyd and heklyd and at the laste sponne," of the bleaching, and finally of its many uses for making clothing, and for sails, and fish nets, and thread, and ropes, and strings ("for bows"), and measuring lines, and sheets ("to reste in"), and sackes, and bagges, and purses (" to put and to kepe thynges in "). Of the making of tow " uneven and full of knobs," used for stuffing into the cracks in ships, and " for bonds and byndynges and matches for candelles, for it is full drye and takyth sone fyre and brenneth." " And so," he concludes somewhat breathlessly, " none herbe is so nedefull to so many dyurrse uses to mankynde as is the flexe." Of the vineyard " closyd about wyth walles and wyth hegges, with a wayte [watch] set in an hyghe place to kepe the vynyerde that the fruyte be not dystroyed." Of the desolation of the vineyard in winter, " but in harueste tyme many comyth and haunteth the vynyerde." Of the delicious smell of a vineyard. Of the damage done by foxes and swine and " tame hounds." " A few hounds," Bartholomew tells us, " wasten and dystroye moo grapes that cometh and eteth therof theuylly [thievishly]." " A vineyard," he concludes, " maye not be kepte nother sauyd but by his socour and helpe that all thynge hath and possesseth in his power and myghte. And kepyth and sauyth all lordly and myghtily." And is there any other writer who in so few words tells us of the woods in those days? Of the " beestis and foulis " therein as well as the herbs, of the woods in summer-time, of the hunting therein, of the robbers and the difficulty of finding one's way? Of the birds and the bees and the wild honey and the delicious coolness of the deep shade in summer, and the " wery wayfarynge trauelynge men "? And the final brief suggestion of the time when forests were veritable boundaries? I believe also that this is the only book in which we are told of the interesting old custom of tying knots to the trees " in token and marke of ye highe waye," and of robbers deliberately removing them. The picture is so perfect that I give it in full :—

" Woods ben wide places wast and desolate y[t] many trees growe in w[t]oute fruyte and also few hauyinge fruyte. And those trees whyche ben bareyne and beereth noo manere fruyte alwaye ben generally more and hygher thañe y[t] wyth fruyte, fewe out taken as Oke and Beche. In thyse wodes ben ofte wylde beestes and foulis. Therein growyth herbes, grasse, lees and pasture, and namely medycynall herbes in wodes foũde. In somer wodes ben bewtyed [beautied] wyth bowes and braunches, w[t] herbes and grasse. In wode is place of disceyte [deceit] and of huntynge. For therin wylde beest ben hunted: and watches and disceytes [deceits] ben ordenyd and lette of houndes and of hunters. There is place of hidynge and of lurkyng. For ofte in wodes theuys ben hyd, and oft in their awaytes and disceytes passyng men cometh and ben spoylled and robbed and ofte slayne. And soo for many and dyuerse wayes and uncerten strange men ofte erre and goo out of the waye. And take uncerten waye and the waye that is unknowen before the waye that is knowen and come oft to the place these theues lye in awayte and not wythout peryll. Therefore ben ofte knottes made on trees and in busshes in bowes and in braunches of trees; in token and marke of ye highe waye; to shewe the certen and sure waye to wayefareynge men. But oft theuys in tornynge and metyng of wayes chaunge suche knottes and signes and begyle many men and brynge them out of the ryght waye by false tokens and sygnes. Byrdes, foules and bein [bees] fleeth to wode, byrdes to make nestes and bein [bees] to gadre hony. Byrdes to kepe themself from foulers and bein [bees] to hyde themself to make honycombes preuely in holowe trees and stockes. Also wodes for thyknesse of trees ben colde with shadowe. And in hete of the sonne wery wayfarynge and trauelynge men haue lykynge to have reste and to hele themself in the shadow. Many wodes ben betwyne dyuers coũtrees and londes: and departyth theym asondre. And by weuynge and castyng togyder of trees often men kepeth and defendyth themself from enymies." [1]

[1] Under "Birch" there is another touch of life in the woods in the Middle Ages. "Wylde men of wodes and forestes useth that sede instede of breede [bread]. And this tree hath moche soure juys and somwhat bytynge. And men useth therfore in spryngynge tyme and in haruest to slyt the ryndes and

Bartholomew's book on herbs ends thus: "And here we shall fynysshe and ende in treatyng of the XVII boke whyche hath treated as ye may openly knowe of suche thynges as the Maker of all thyng hath ordered and brought forth by his myghty power to embelyssh and araye the erthe wyth and most specyally for ye fode of man and beast."

At the end of the book is the poem which has caused so much controversy amongst bibliographers. In this Wynken de Worde definitely states that Caxton had a share in the first printing of this book at Cologne:—

> "And also of your charyte call to remembraunce
> The soule of William Caxton first pry̅ter of this boke.
> In laten tonge at Coleyn hyself to auauce
> That every well disposed man may therein loke."

In spite of this, modern bibliographers are of opinion that Caxton could not have played even a subordinate part in the printing of this book at Cologne.

De Worde also refers to the maker of the paper [1]:—

> ". . . John Tate the yonger. . . .
> Which late hathe in England doo make this paper thynne
> That now in our Englysh this boke is prynted Inne."

There is charm as well as pathos in the verses on the reproduction of manuscripts in book form, showing us vividly what the recent discovery of the art of printing meant to the scholars of that day. The simile of Phœbus "repairing" the moon is very apt.

> "For yf one thyng myght laste a M yere
> Full sone comyth aege that frettyth all away;
> But like as Phebus wyth his bemes clere
> The mone repeyreth as bryght as ony day
> Whan she is wasted ryght; so may we say
> Thise bokes old and blynde whan we renewe
> By goodly pryntyng they ben bryght of hewe."

---

to gader ye humour that comyth oute therof and drynkyth in stede of wyn. And such drynke quencheth thurste. But it fedyth not nother nourryssheth not, nother makyth men dronke."

[1] In regard to this paper (probably the first made in England for printing) see Bibliography, p. 204.

The last verse of the poem is as follows :—

> " Nowe gloryous god that regnest one in thre
> And thre in one graunte vertu myght and grace
> Unto the prynter of this werke that he
> May be rewarded in thy heuenly place
> And whan the worlde shall come before thy face
> There to receue accordyng to desert
> Of grace and mercy make hym then expert."

The treatise on herbs formed, as we have seen, only a part of Bartholomew's *De Proprietatibus Rerum*, and, to speak strictly, the first printed English herbal was the small quarto volume published by Richard Banckes in 1525. It was the beginning of a series of small books [1] chiefly in black letter. All of them, though issued from different presses, have nearly the same title, and they vary only slightly from the original *Banckes's Herbal*. The title of this Herbal is—

" Here begynneth a new mater / the whiche sheweth and | treateth of ye vertues & proprytes of her- | bes / the whiche is called | an Herball ·.· | ℭ Cum gratia & priuilegio | a rege indulto |
(*Colophon*) ℭ Imprynted by me Rycharde Banckes / dwellynge in | Lōdō / a lytel fro ye Stockes in ye Pultry / ye XXV day of | Marche. The yere of our Lorde MCCCCC. & XXV."

We do not know who the author of this book was, and it has been suggested that it is based on some mediæval English manuscript now lost. Certainly when one reads this anonymous work known as *Banckes's Herbal* one is struck not only by its superiority to the later and more famous *Grete Herball*, but also by its greater charm. It gives the impression of being a compilation from various sources, the author having made his own selection from what pleased him most in the older English manuscript herbals. It seems to have been a labour of love, whereas the *Grete Herball* is merely a translation. It is almost certain that the writer made use of one of the numerous

[1] For dates, full titles, etc., of all the editions of *Banckes's Herbal* see Bibliography of English Herbals.

manuscript versions of Macer's Herbal, which in parts *Banckes's Herbal* resembles very closely, and the chapter on rosemary shows that he had access to one of the copies of the manuscript on the virtues of rosemary which was sent by the Countess of Hainault to Queen Philippa. He does not give the beautiful old tradition preserved in that manuscript,[1] but he ascribes wonderful virtues to this herb, with the same loving enthusiasm and almost in the same words. Of rosemary in *Banckes's Herbal* we read :—

" Take the flowers thereof and make powder thereof and binde it to thy right arme in a linnen cloath and it shale make theee light and merrie.

" Take the flowers and put them in thy chest among thy clothes or among thy Bookes and Mothes shall not destroy them.

" Boyle the leaves in white wine and washe thy face there-with and thy browes and thou shalt have a faire face.

" Also put the leaves under thy bedde and thou shalt be delivered of all evill dreames.

" Take the leaves and put them into wine and it shall keep the wine from all sourness and evill savours and if thou wilt sell thy wine thou shalt have goode speede.

" Also if thou be feeble boyle the leaves in cleane water and washe thyself and thou shalt wax shiny.

" Also if thou have lost appetite of eating boyle well these leaves in cleane water and when the water is colde put there-unto as much of white wine and then make sops, eat them thereof wel and thou shalt restore thy appetite againe.

" If thy legges be blowen with gowte boyle the leaves in water and binde them in a linnen cloath and winde it about thy legges and it shall do thee much good.

" If thou have a cough drink the water of the leaves boyld in white wine and ye shall be whole.

[1] See p. 44.

INITIAL LETTERS FROM "BANCKES'S HERBAL"

"Take the Timber thereof and burn it to coales and make powder thereof and rubbe thy teeth thereof and it shall keep thy teeth from all evils. Smell it oft and it shall keep thee youngly.

"Also if a man have lost his smellyng of the ayre that he may not draw his breath make a fire of the wood and bake his bread therewith, eate it and it shall keepe him well.

"Make thee a box of the wood of rosemary and smell to it and it shall preserve thy youth."

That *Banckes's Herbal* achieved immediate popularity is attested by the fact that the following year another edition of it was issued, and during the next thirty years various London printers issued the same book under different titles.[1] Robert Wyer[2] ascribed the authorship of those he issued to Macer, and in the edition of 1530 he added, after "Macer's Herbal," "Practysed by Dr. Lynacro." Whether this statement is true it is impossible to discover, but we know that the great doctor died some years before Wyer set up as a printer, and his name does not appear in any of the subsequent editions of the herbal issued by other printers. In Wyer's edition there are some good initial letters very similar to those used by Wynkyn de Worde.

The most interesting edition of the herbal is that printed by William Copland, in which first appear the additional chapters on "The virtues of waters stylled," "The tyme of gathering of sedes" and "A general rule of all maner of herbes." He issued two editions bearing the same title and differing only

[1] See Bibliography of English Herbals.

[2] Robert Wyer was one of the most famous printers of the early sixteenth century. He came of a Buckinghamshire family and was probably a near relation of John Wyer, also a printer who lived in Fleet Street, for both of them used the device of St. John the Evangelist. He served his apprenticeship to Richard Pynson, whose printing press was in the rentals of Norwich House near the site of the present Villiers Street, and on Pynson's death succeeded to the business. In both his editions of the herbal there is his well-known device of St. John the Evangelist bareheaded and dressed in a flowing robe, sitting under a tree on an island and writing on a scroll spread over his right knee. At his right hand is an eagle with outstretched wings holding an ink-well in its beak, and in the background are the towers and spires of a great city.

in the woodcuts and the colophon. The title is "A boke of the | propreties of Herbes called an her- | ball, whereunto is added the tyme y^e | herbes, floures and Sedes shold | be gathered to be kept the whole, ye- | re, with the vertue of ye Herbes whē | they are stylled. Al- | so a generall rule of all ma- | ner of Herbes drawen | out of an auncyent | booke of Phisyck | by W. C." The woodcut in the first edition is three "Tudor" roses in a double circle with a crown over one of the roses and across the riband "Kȳge of floures." In the second edition the woodcut is a quaint little representation of a lady seated in a garden. One man standing behind her is holding her and another is walking towards her. The three figures are near a wall, on the other side of which several men are apparently conversing. Who W. C. was is uncertain. In the *Dictionary of National Biography* William Copland is said to be both the author and the printer of the book, but in many catalogues (notably in that of the British Museum) Walter Cary figures as the author. In a lengthy account of the Carys in *Notes and Queries* (March 29, 1913) Mr. A. L. Humphreys disposes conclusively of the supposition that W. C. can stand for Walter Cary.

"*A Boke of the Properties of Herbes* bears on the title-page the initials W. C., which may stand either for Copland or Cary. This was one of several editions of *Banckes's Herbal,* then very popular, and although it may have been edited or promoted in some way by a Walter Cary, it could not have been by the one who wrote *The Hammer for the Stone.* The 'Herball' was issued somewhere about 1550 and various editions of it exist, but all these appeared when the Walter Cary we are considering was a child. There is, however, a connection between the Carys and herbals, because it is well known that Henry Lyte (1529–1607) of Lytes Cary was the famous translator of Dodoens's *Herball* (1578), and he had a herbal garden at Lytes Cary."

Ames in his *Typographical Antiquities* describes the two editions, which are identical, as though they were two different books, and ascribes one to Walter Cary and the other to William Copland. We have only Ames's authority for the supposition that Copland was the compiler as well as the printer. The herbal in question is merely another edition of *Banckes's Herbal*, but it is quite possible that the three additional chapters at the end were "drawen out of an auncyent booke of Physick" by Copland.[1]

Two editions of *Banckes's Herbal* are ascribed, on account of the wording of the title, to Antony Askham, and the title is so attractive that it is a disappointment to find that the astrological additions "declaryng what herbes hath influence of certain sterres and constellations," etc., do not appear in any known copy of the herbal. This astrological lore from the famous man who combined the professions of priest, physician and astrologer in the reign of Edward VI. would be of remarkable interest. But it has been pointed out by Mr. H. M. Barlow[2] that, if the bibliographers who have attributed the work to Askham had examined the title of the work with greater care, they would have observed that the phrase "by Anthonye Askham" refers not to the substance of the book itself (which is merely another edition of *Banckes's Herbal*) but to the "Almanacke" from which the additions were intended to be taken, though apparently they were never printed. The title of "Askham's" Herbal is—

"A lytel | herball of the | properties of her- | bes newely amended and corrected, | with certayne addicions at the ende | of the boke, declarying what herbes | hath influence of certaine Sterres | and constellations, whereby may be | chosen the beast and most luckye | tymes and dayes of their mini- | stracion, accordyinge to the | Moone being in the sig- | nes of heauen, the | which is dayly | appoynted | in the | Almanacke; made and

[1] Ames catalogues two other editions of the herbal by "W. C.," one published by Anthony Kitson and the other by Richard Kele, but no known copies of these exist.

[2] Proceedings of the Royal Society of Medicine, 1913.

gathered | in the yere of our Lorde god | M.D.L. the XII. day of Fe- | bruary by Anthonye | Askham Phi- | sycyon.

" (*Colophon.*) Imprynted at | London in Flete- | strete at the signe of the George | next to Saynte Dunstones | Churche by Wylly- | am Powell. | In the yeare of oure Lorde | M.D.L. the twelfe day of Marche."

There are some charming prescriptions to be found in "Askham's" Herbal. Under "rose," for instance, we have recipes for "melroset," "sugar roset," "syrope of Rooses," "oyle of roses" and "rose water."

" Melrosette is made thus. Take faire purified honye and new read rooses, the whyte endes of them clypped awaye, thā chop theym smal and put thē into the Hony and boyl thē menely together; to know whan it is boyled ynoughe, ye shal know it by the swete odour and the colour read. Fyve yeares he may be kept in his vertue; by the Roses he hath vertue of comfortinge and by the hony he hath vertu of clensinge.

" Syrope of Rooses is made thus. Some do take roses dyght as it is sayd and boyle them in water and in the water strayned thei put suger and make a sirope thereof; and some do make it better, for they put roses in a vessell, hauing a strayght mouthe, and they put to the roses hote water and thei let it stande a day and a night and of that water, putting to it suger, thei do make sirope, and some doe put more of Roses in the forsaid vessel and more of hote water, and let it stande as is beforesaide, and so they make a read water and make the rose syrope. And some do stāpe new Roses and then strayne out the joyce of it and suger therwyth, they make sirope : and this is the best making of sirope. In Wynter and in Somer it maye be geuen competently to feble sicke melācoly and colorike people.

" Sugar Roset is made thus—Take newe gathered roses and stāpe them righte smal with sugar, thā put in a glasse XXX. dayes, let it stande in ye sunne and stirre it wel, and medle it well together so it may be kept three yeares in his vertue. The quātitie of sugar and roses should be thus. In IIII. pound of sugar a pounde of roses.

" Oyle of roses is made thus. Some boyle roses in oyle and kepe it, some do fyll a glasse with roses and oyle and they boyle it in a caudron full of water and this oyle is good. Some stampe fresh roses with oyle and they put it in a vessel of glasse and set it in the sūne IIII. dais and this oyle is good.

" Rose water. Some do put rose water in a glass and they put roses with their dew therto and they make it to boile in water thā thei set it in the sune tyll it be readde and this water is beste."

Under the same flower we find this fragrant example of the widespread mediæval belief in the efficacy of good smells :—

" Also drye roses put to ye nose to smell do cōforte the braine and the harte and quencheth sprite."

The herbalists were never weary of teaching the value of sweet scents.[1] " If odours may worke satisfaction," wrote Gerard in his *Herball,* " they are so soveraigne in plants and so comfortable that no confection of the apothecaries can equall their excellent vertue." One of the most delicious " scent " prescriptions in Askham is to be found under Violet—" For thē that may not slepe for sickness seeth this herb in water and at euen let him soke well hys feete in the water to the ancles, whā he goeth to bed, bind of this herbe to his temples and he shall slepe wel by the grace of God."

The most curious recipe is that under " woodbinde." " Go to the roote of woodbinde and make a hole in the middes of the roote, than cover it well againe y[t] no ayre go out nor that

[1] The popular belief in the power of sweet-smelling herbs to ward off infection of the much-dreaded plague rose to its height in Charles II.'s reign, when bunches of rosemary were sold for six and eightpence. Till recently there were at least two survivals of this belief in herbal scents—the doctor's gold-headed cane (formerly a pomander carried at the end of a cane) and the little bouquets carried by the clergy at the distribution of the Maundy Money in Westminster Abbey.

no rayne go in, no water, nor earth nor the sune come not to much to it, let it stande so a night and a day, thā after that go to it and thou shalt fynde therein a certayne lycoure. Take out that lycoure with a spone and put it into a clean glas and do so every day as long as thou fyndest ought in the hole, and this must be done in the moneth of April or Maye, than anoynt the sore therwith against the fyre, thā wete a lynnen clothe in the same lycoure and lappe it about the sore and it shal be hole in shorte space on warrantyse by the Grace of God."

Unlike the later *Grete Herball,* Askham gives some descriptions of the herbs themselves, notably in the case of alleluia (wood-sorrel), water crowfoot, and asterion.

"This herbe alleluia mē call it Wodsour or Stubwort, this herbe hath thre leaves ye which be roūd a litel departed aboue and it hath a whyte flour, but it hath no lōge stalkes and it is Woodsoure and it is like thre leued grasse. The vertue of this herbe is thus, if it be rosted in the ashes in red docke leaves or in red wort leaves it fretteth awai dead flesh of a wounde. This herbe groweth much in woodes."

Water crowfoot: "This herb that men call water crowfoot hath yelow floures, as hath crowfoot and of the same shap, but the leves are more departed as it were Rammes fete, and it hath a long stalke and out of that one stalke groweth many stalkes smal by ye sides. This herb groweth in watery places."

"Asterion or Lunary groweth among stoones and in high places, this herb shyneth by night and he bringeth forth purple floures hole and rounde as a knockebell or else lyke to foxgloves, the leves of this herbe be rounde and blew and they have the mark of the Moone in the myddes as it were thre leved grasse, but the leaves therof be more and they be round as a peny. And the stalk of this herb is red and thyse herb semeth as it were musk and the joyce therof is yelow and this groweth in the new Moone without leve and euery day spryngeth a newe leaue to the ende of fyftene dayes and after fyftene dayes it looseth euery day a leaue as the Moone waneth and it springeth and waneth as doth the Moone and where that it groweth there groweth great quantitie.

"The vertue of this herbe is thus—thei that eat of the beris or of the herbe in waning of the moone, whā he is in signo virginis if he have the falling euell he shal be hole thereof or if he beare thys about his neck he shal be holpen without doute. And it hath many more vertues than I can tell at this tyme."

One of the unidentified herbs is called "sene," and we are given the somewhat vague geographical information, "It groweth in the other syde the sea and moste aboute Babilon."

Another small book printed by William Copland must be mentioned, for, although it is not a herbal, it contains a great deal of curious herb lore not to be found elsewhere. This is *The boke of secretes of Albartus Magnus of the vertues of Herbes, Stones, and certaine beastes.* Who the author was is unknown, but he was certainly not Albert of Bollstadt (1193–1280), Bishop of Ratisbon, the scholastic philosopher to whom it was ascribed, probably in order to increase its sale. There is one philosophical remark which is not unworthy of the famous Bishop: "Every man despiseth ye thyng whereof he knoweth nothynge and that hath done no pleasure to him." But for the most part it deals with the popular beliefs concerning the mystical properties of herbs, stones and animals.

Of celandine the writer tells us: "This hearbe springeth in the time in ye which the swallowes and also ye Eagles maketh theyr nestes. If any man shal have this herbe with ye harte of a Molle (mole) he shall overcome all his enemies. . . . And if the before named hearbe be put upon the headde of a sycke man if he should dye he shal syng anone with a loud voyce, if not he shall weep."

"Perwynke when it is beatē unto pouder with wormes of ye earth wrapped aboute it and with an herbe called houslyke it induceth love between man and wyfe if it bee used in their meales . . . if the sayde confection be put in the fyre it shall be turned anone unto blue coloure."

Of the herb which, he tells us, "the men of Chaldea called roybra, he says: "He that holdeth this herbe in hys hāde with an herbe called Mylfoyle or yarowe or noseblede is sure from all feare and fantasye or vysion. And yf it be put with the juyce of houselyke and the bearers hands be anoynted with it and the residue be put in water if he entre in ye water where fyshes be they wil gather together to hys handes . . . and if hys hande be drawē forth they will leape agayne to theyre owne places where they were before."

Of hound's tongue: "If ye shall have the aforenamed herbe under thy formost toe al the dogges shall kepe silence and shall not have power to bark. And if thou shalt put the aforesayde thinge in the necke of any dogge so yt he maye not touche it with his mouthe he shalbe turned always round about lyke a turning whele untill he fall unto the grounde as dead and this hath bene proved in our tyme."

Of centaury: "If it be joyned with the bloude of a female lapwing or black plover and be put with oyle in a lampe, all they that compasse it aboute shal beleue themselves to be witches so that one shall beleve of an other that his head is in heaven and his fete in the earth. And if the aforesaid thynge be put in the fire whan the starres shine it shall appeare yt the sterres runne one agaynste another and fyght."

Of vervain: "This herbe (as witches say) gathered, the sunne beyng in the signe of the Ram, and put with grayne or corne of pyonie of one yeare olde healeth them yt be sicke of ye falling sykenes."

Of powder of roses: "If the aforesayde poulder be put in a lampe and after be kindled all men shall appeare blacke as the deuell. And if the aforesaid poulder be mixed with oyle of the olyue tree and with quycke brymstone and the house anointed wyth it, the Sunne shyning, it shall appeare all inflamed."

WOODCUT FROM THE TITLE-PAGE OF THE "GRETE HERBALL" (1526)

Of verbena: "Infants bearing it shalbe very apte to learne and louing learnynge and they shalbe glad and joyous."

It is the only book on the virtues of herbs in which I have found a recipe to revive drowning flies and bees! This is to be done by placing them in warm ashes of pennyroyal, and then "they shall recover their lyfe after a little tyme as by ye space of one houre." The book ends with a curious philosophical dissertation, "Of the mervels of the worlde," which is followed by a series of charms—to stop a cock crowing, to make men look as though they had no heads, to obtain rule over all birds, to keep flies away from a house, to write letters which can only be read at night, to make men look as though they had "the countenance of a dog," to make men seem as though they had three heads, to understand the language of birds, to make men seem like angels, and to put things in the fire without their being consumed.

Though lacking in the charm of the quaint and typically English *Banckes's Herbal*, the most famous of the early printed herbals was the *Grete Herball* printed by Peter Treveris in 1526.[1]

"The grete herball | whiche geueth parfyt knowlege and under- | standyng of all maner of herbes & there gracyous vertues whiche god hath | ordeyned for our prosperous welfare and helth, for they hele & cure all maner | of dyseases and sekenesses that fall or mysfortune to all maner of creatoures | of god created, practysed by many expert and wyse maysters, as Auicenna and | other &c. Also it geueth full parfyte under-standynge of the booke lately pryn | ted by me (Peter treveris) named the noble experiens of the vertuous hand | warke of surgery."

(*Colophon.*) "Imprentyd at London in South- | warke by me peter Treueris, dwel- | lynge in the sygne of the wodows | In the yere of our Lorde god M.D. | XXVI the XXVII day of July."

According to the introduction it was compiled from the

[1] For dates of later editions see Bibliography of English Herbals.

works of "many noble doctoures and experte maysters in medecines, as Auicenna, Pandecta, Constantinus, Wilhelmus, Platearius, Rabbi Moyses, Johannes Mesue, Haly, Albertus, Bartholomeus and more other." But with the exception of the preface the *Grete Herball* is a translation of the well-known French herbal, *Le Grant Herbier*. Until about 1886 *Le Grant Herbier* was supposed to be a translation of the *Herbarius zu Teutsch*, published at Mainz in 1485, or of the *Ortus Sanitatis*, printed also at Mainz in 1491. The *Herbarius zu Teutsch*, which was probably compiled by a Frankfort physician, is a fine herbal beautifully illustrated, and the later *Ortus Sanitatis* is by some authorities supposed to be a Latin translation of it. To judge from the preface to the German Herbarius it was a labour of love, undertaken by a man who apparently was possessed of ample wealth and leisure; for in his preface he tells us that he "caused this praiseworthy work to be begun by a Master learned in physic," and then, finding that as many of the herbs did not grow in his native land he could not draw them "with their true colours and form," he left the work unfinished and journeyed through many lands—Italy, Croatia, Albania, Dalmatia, Greece, Corfu, Candia, Rhodes, Cyprus, the Holy Land, Arabia, Babylonia and Egypt. He was accompanied by "a painter ready of wit and cunning and subtle of hand," and was thus able to have the herbs "truly drawn." The book he compiled on his return was long regarded as the original of the French herbal, *Le Grant Herbier*, but in 1866 Professor Giulio Camus found two fifteenth-century manuscripts in the Biblioteca Estense at Modena, one the Latin work commonly known from the opening words as *Circa Instans*, and the other a French translation of the same manuscript. It was always supposed by medical historians that the *Circa Instans* was written by Matthaeus Platearius of Salerno in the twelfth century, but in Professor Camus's memoir, *L'Opéra Saleritana "Circa Instans" ed il testo primitivo del*

[1] For fuller bibliographical details of the *Herbarius zu Teutsch* and the *Ortus Sanitatis* see Bibliography of Foreign Herbals.

"*Grand Herbier in Francoys*" *secundo duo codici del secolo XV conservati nella Regia Biblioteca Estense*, there are reproduced the French verses in which occurs the line, "Il a esté escript Millccc cinquante et huit," and Mr. H. B. Barlow[1] supports the deduction that *Circa Instans* was not written by a Salernitan physician, but by a writer described in the verses as "Bartholomaeus minid' senis" in 1458. *Le Grant Herbier*, of which the English *Grete Herball* is a translation, is a version of the French manuscript translation of *Circa Instans*, and therefore, as *Circa Instans* is older than either the *Herbarius zu Teutsch* or the Latin *Ortus Sanitatis*, it would seem that it is the real original of our *Grete Herball*. The preface to the *Grete Herball*, however, bears a strong resemblance to that of the German Herbarius, of which I quote a part from Dr. Arber's translation, made from the second (Augsburg) edition of 1485. They have been placed in parallel columns to show how closely the English preface follows that of the German Herbarius.

Preface to the *Herbarius zu Teutsch*.

"Many a time and oft have I contemplated inwardly the wondrous works of the creator of the universe: how in the beginning He formed the heavens and adorned them with goodly shining stars, to which he gave power and might to influence everything under heaven. Also how he afterwards formed the four elements: fire, hot and dry—air, hot and moist—water, cold and moist—earth, dry and cold—and gave to each a nature of its own; and how after this the same Great Master of Nature made and formed herbs of many sorts and animals of all kinds and last of all Man, the noblest of all created things. Thereupon I thought on the wondrous order which

Preface to *The Grete Herball*.

"Consyderynge the grete goodnesse of almyghty God creatour of heven and erthe, and al thynge therin comprehended to whom be eternall laude and prays etc. Consyderynge the cours and nature of the foure elementes and qualytees where to ye nature of man is inclyned, out of the whiche elementes issueth dyvers qualytees infyrmytees and dyseases in the corporate body of man, but god of his goodnesse that is creatour of all thynges hath ordeyned for mankynd (whiche he hath created to his own lykenesse)

[1] Proceedings of the Royal Society of Medicine.

the Creator gave these same creatures of His, so that everything which has its being under heaven receives it from the stars and keeps it by their help. I considered further how that in everything which arises, grows, lives or soars in the four elements named, be it metal, stone, herb or animal, the four natures of the elements, heat, cold, moistness and dryness, are mingled. It is also to be noted that the four natures in question are also mixed and blended in the human body in a measure and temperament suitable to the life and nature of man; while man keeps within this measure . . . he is strong and healthy, but as soon as he steps or falls beyond . . . which happens when heat takes the upper hand and strives to stifle cold or on the contrary when cold begins to suppress heat . . . he falls of necessity into sickness and draws nigh unto death. . . . Of a truth I would as soon count the leaves on the trees or the grains of sand in the sea as the things which are the causes of man's sickness. It is for this reason that so many thousands and thousands of perils and dangers beset man. He is not fully sure of his health or his life for one moment. While considering these matters, I also remembered how the Creator of Nature, who has placed us amid such dangers has mercifully provided us with a remedy, that is with all kinds of herbs, animals and other created things. . . . By virtue of these herbs and created things the sick man may recover the temperament of the four elements and the health of his body. Since then man can have no greater nor nobler treasure on earth than bodily health, I came to the conclusion that I could not perform any more useful and holy work than to

for the grete and tender love, which he hath unto hym, to whom all thinges erthely he hath ordeyned to be obeysant, for the sustentacyon and helthe of his lovynge creature mankynde whiche is onely made egally of the foure elementes and qualitees of the same, and when any of these foure habounde or hath more domynacyon, the one than the other it constrayneth ye body of man to grete infyrmytees or dyseases, for the which ye eternall god hath gyven of his haboundante Grace, vertues in all maner of herbes to cure and heale all maner of sekenesses or infyrmytees to hym befallying through the influent course of the foure elementes beforesayd and of the corrupcyons and ye venymous ayres contrarye ye helthe of man. Also of onholsam meates or drynkes, or holsam meates or drynkes taken ontemperatly whiche be called surfetes that bryngeth a man sone to grete dyseases or sekenesse, whiche dyseases ben of nombre and ompossoyble to be rehersed, and fortune as well in vilages where as nother surgeons nor phisicians be dwellyng nygh by many a myle, as it dooth in good townes where they be redy at hande, wherefore brotherly love compelleth me to wryte thrugh ye gyftes of the holy ghost shewynge and enformynge how man may be holpen with grene herbes of the

compile a book in which could be contained the virtue and nature of many herbs and other created things, together with their true colours and for the help of all the world, and the common good, therefore I caused this praiseworthy work to be begun by a Master learned in physic who, at my request gathered into a book the nature and virtue of many herbs out of the acknowledged masters of physic, Galen, Avicenna, Serapio, Dioscorides, Pandectarius, Platearius and others."

gardyn and wedys of ye feldys as well as by costly receptes of the potycarys prepayred."

The illustrations in the *Grete Herball* are poor, being merely inferior copies of those in the later editions of the *Herbarius zu Teutsch*.[1] In the majority of cases it is impossible to identify the plant from the figure, and the same figure is sometimes prefixed to different plants. But if the illustrations are poor and dull the frontispiece and the full-page woodcut of the printer's mark are very much the reverse. The frontispiece is a charming woodcut of a man holding a spade in his right hand and gathering grapes, and a woman throwing flowers and herbs out of her apron into a basket. There are two figures in the lower corners, the one of a male and the other of a female mandrake. The woodcut of the printer's mark at the end sheds an interesting ray of light on the Peter Treveris who issued the two first editions of this Herball.[2] The woodcut represents two wodows [3] (savages), a man and a woman, on either side of a tree, from which is suspended a shield with Peter Treveris's

[1] The illustrations in the second and later editions of the *Herbarius zu Teutsch* are very inferior to those in the first, which are beautiful. *The vertuose boke of Distillacyon of the waters of all maner of Herbes* (1527), translated by Laurence Andrew from the *Liber de arte distillandi* by Hieronymus Braunschweig, is illustrated with cuts from the same wood-blocks as the *Grete Herball*.

[2] Titles and dates of the subsequent editions issued by Thomas Gibson (1539) and Jhon Kynge (1561) will be found in the Bibliography of English Herbals.

[3] Treveris had his printing office in Southwark, at the sign of the "Wodows."

initials. Ames supposes that Treveris was a native of Trèves and took his name from that city, but it is more likely that he was a member of the Cornish family of Treffry, which is sometimes spelt Treveris. A Sir John Treffry, who fought at Poitiers, took as supporters to his arms a wild man and woman, and one likes to find that one of his descendants perpetuated the memory of his gallant ancestor by adopting the same sign for his trade device.

The *Grete Herball* is alphabetically arranged, for the idea of the natural relationship of plants was unknown at that time. But we find a " classification " of fungi. " Fungi ben musherons. There be two maners of them, one maner is deadly and sleeth them that eateth of them and the other dooth not " ! As in most sixteenth- and seventeenth-century herbals, there are quaint descriptions of a good many things besides herbs. The most gruesome of these is a substance briefly described as " mummy," and the accompanying illustration is of a man digging beside a tomb. " Mummy," one reads, " is a maner of spyces or confectyons that is founde in the sepulchres or tombes of dead bodyes that haue be confyct with spyces. And it is to wyte that in olde tyme men were wont to confyct the deed corpses and anoynte them with bawme and myre smellynge swete. And yet ye paynims about babylon kepe that custome for there is grete quantity of bawme. And this mummye is specially founde about the brayne and about the maronge in the rydge bone. For the blode by reason of the bawme draweth to the brayne and thereabout is chauffed. And lykewise is the brayne brent and parched and is the quantyte of mommye and so the blode is mroeued in the rydge of the backe. That mommye is to be chosen that is bryght blacke stynkynge and styffe. And that y^t is whyt and draweth to a dymme colour and that is not stynkynge nor styffe, and that powdreth lightly is naught. It hath vertue to restrayne or staunche." [1]

[1] The use of " mummy " is not only mentioned by all the later herbalists up to the end of the seventeenth century, but is even to be found in MS. still-

WOODCUT OF PETER TREVERIS' SIGN OF THE "WODOWS" FROM THE "GRETE HERBALL" (1529)

WOODCUT FROM THE TITLE-PAGE OF THE FOURTH EDITION OF THE "GRETE HERBALL" (1561)

Other substances described are salt, cheese, pitch, lead, silver, gold, amber, water, starch, vinegar, butter, honey and the lodestone. The dissertation on water shows very clearly that our ancestors regarded bathing as a fad, and a dangerous fad at that. The writer gloomily observes, " many folke that hath bathed them in colde water haue dyed or they came home." And those who are foolish enough to drink water he warns by quoting the authority of " Mayster Isaac," who " sayth that it is impossible for them that drynketh overmuche water in theyr youth to come to ye age that God hath ordeyned them." In the description of the lodestone we find the well-known popular belief about ships being drawn to their destruction. " The lodestone, the adamant stone that draweth yren hath myghte to draw yren as Aristotle sayth. And is founde in the brymmes of the occyan see. And there be hillis of it and these hyllis drawe ye shippes that haue nayles of yren to them and breke the shyppes by drawynge of the nayles out." The accompanying illustration is of a sinking ship with a man going towards the hill of adamant with uplifted hands, while another man is swimming, and a third sits calmly in the ship.

In view of the free use of honey in olden times, the account of honey in the *Grete Herball* seems inadequate. " Hony is made by artyfyce and craft of bees. The whyche bees draweth the thynnest parte of the floures and partelye of the thickest and moost grosse and thereof maketh hony and waxe and also they make a substaunce that is called the honycombe. The tame hony is that that is made in the hous or hyues that labourers ordeyneth for the sayd bees to lodge and worke in. Hony is whyte in cold places and browne in warm place. And hony ought to be put in medicyne and may be kept C yeeres. There

room books. In the Fairfax still-room book a recipe for wounds said to have been procured from " Rodolphus Goclerius, professor of Phisicke in Wittenburghe," begins thus : " Take of the moss of a strangled man 2 ounces, of the mumia of man's blood, one ounce and a halfe of earth-worms washed in water or wine and dyed," etc.

is an other that is called wylde hony and is found in woodes and is not so good as the other and is more bytter. Also there is a honey called castanea because it is made of chestayne floures that the bees sucketh and is bytter."

In the *Grete Herball*, as in *Banckes's Herball*, we find numerous instances of the use of herbs as amulets or for their effect on the mind, and for the smoking of patients with their fumes. I quote the following :—

"Betony. For them that be ferfull. For them that ben to ferfull gyue two dragmes of powdre hereof wt warme water and as moche wyne at the tyme that the fere cometh."

"Buglos. To preserve the mynde. This herbe often eaten confermeth and conserueth the mynde as many wyse maysters sayth."

"To make folke mery. Take the water that buglos hath bē soden in and sprynkle it about the hous or chambre and all that be therein shall be mery."

"Vervain. To make folke mery at ye table. To make all them in a hous to be mery take foure leaves and foure rotes of vervayn in wyne, than sprycle the wine all about the hous where the eatynge is and they shall be all mery."

"Musk. Agaynst weyknesse of the brayne smel to musk."

"Struciūn. Against lytargye blowe the powdre of the sede in to the nose or elles sethe the sede thereof and juice of rue in stronge vyneygre and rubbe the hynder parte of ye head therwith."

"Artemisia. To make a child mery hange a bondell of mugwort or make smoke thereof under the chylde's bedde for it taketh away annoy for hem."

"Rosemary. For weyknesse of ye brayne. Agaynst weyknesse of the brayne and coldenesse thereof, sethe rosmarin in wyne and lete the pacyent receye the smoke at his nose and kepe his heed warme."

"Southernwood. The fume of it expelleth all serpents out of the house and what so ever there abydeth dyeth."

There are two delicious violet recipes for "Syrope of Vyolettes" and "oyle of vyolettes."

"Syrope of vyolettes ī made in this maner—Sethe vyolettes in water and lete it lye all nyght in ye same water. Than poure and streyne out the water, and in the same put sugre and make your syrope.

"Oyle of vyolettes is made thus. Sethe vyolettes in oyle and streyne it. It will be oyle of vyolettes."

It is in this herbal that we find the first avowal of disbelief in the supposed powers of the mandrake.

"There be two maners the male and the female, the female hath sharpe leves. Some say that it is better for medycyne than the male but we use of bothe. Some say that the male hath figure of shape of a man. And the female of a woman but that is fals. For Nature never gaue forme or shape of mākynde to an herbe. But it is of troughe that some hath shaped suche fygures by craft as we have fortyme herde say of labourers in the feldes."

The *Grete Herball* ends thus—

"O ye worthy reders or practicyens to whome this noble volume is presēt. I beseche you take intellygence and beholde ye workes and operacyōs of almighty god which hath endewed his symple creature mankynde with the graces of ye holy goost to have parfyte knowlege and understandynge of the vertue of all manner of herbes and trees in this booke comprehendyed and everyche of them chaptred by hymselfe and in every chaptre dyuers clauses where is shewed dyuers maner of medycunes in one herbe comprehended whiche ought to be notyfyed and marked for the helth of man in whom is repended ye hevenly gyftes by the eternall Kynge to whom be laude and prayse everlastynge. Amen."

The only important books Treveris published besides the *Grete Herball* were the two English translations of Hieronymus Braunschweig's works (*The noble experyence of the virtuous Handy-worke of Surgeri* and *The vertuouse Book of the Dystillacion of the Waters of all maner of Herbes*) and the handsome edition of Trevisa's translation of Higden's *Polychronicon*. *The vertuouse Book of the Dystillacion of the Waters of all maner of Herbes* is well printed, but the illustrations are from the same inferior German cuts as those in the *Grete Herball*. The book was translated into English by Laurence Andrew and, though strictly it does not come within the category of herbals, part of the preface is too beautiful to omit. "Lerne the hygh and meruelous vertue of herbes. Knowe how inestimable a preservative to the helth of man god hath provyded growying euery daye at our hande, use the effectes with reverence, and give thankes to the maker celestyall. Beholde how moch it excedeth to use medecyne of efycacye naturall by God ordeyned then wicked wordes or charmes of efycacye unnaturall by the dyuell enuented, whiche yf thou doste well marke, thou shalt have occasyon to gyue the more louynges and praise to oure sauyour, by redynge this boke and knowlegying his benyfites innumerable. To whose prayse, and helthe of all my crysten bretherne, I have taken upon me this symple translacyon, with all humble reverence ever redy to submit me to the correccion of the lerned reder."

CHAPTER III

TURNER'S HERBAL AND THE INFLUENCE OF THE FOREIGN HERBALISTS

" In the beginning of winter the Goldfinches use muche to haunte this herbe [teazle] for the sedes sake whereof they are very desyrous."—*Turner's Herbal*, 1551.

LIKE so many sixteenth-century notabilities, William Turner, commonly known as the father of English botany, was remarkably versatile, for he was a divine, a physician and a botanist. He was a native of Morpeth, Northumberland, and was born in Henry VIII.'s reign : the exact date is unknown. His father is supposed to have been a tanner. We know nothing of his early education, but he entered what is now Pembroke College,[1] Cambridge, under the patronage of Thomas Lord Wentworth. This he himself tells us in the preface to the second part of his herbal, which is dedicated to Lord Wentworth of the next generation. " And who hath deserved better to have my booke of herbs to be given to him, than he, whose father with his yearly exhibition did helpe me, beying student in Cambridge of Physik and philosophy ? Whereby with some further help and study am commed to this pore knowledge of herbes and other simples that I have. Wherefore I dedicate unto you this my litle boke, desyring you to defende it against the envious evil speakers, which can alow nothing but that they do themselves : and the same I give unto your Lordship, beseeching to take it in the stede of a better thyng, and for a token of my good will toward you, and all your father's houshold, which thing if ye

[1] Then " Marie Valence Hall." (Founded in 1347 by Marie widow of Aylmer de Valence, Earl of Pembroke.)

do, as sonne as I shall have convenient lesure, ye shall have the third and last parte of my herball also. Almighty God kepe you and all youres. Amen."

At Cambridge Turner was intimate with Nicholas Ridley (afterwards the famous Bishop of London), and though it is interesting to know that Ridley instructed him in Greek, it is even more attractive to learn that the future bishop also initiated him into the mysteries of tennis and archery. Turner did well at the university, for he was elected Junior Fellow of his college in 1531 and Joint Treasurer in 1532, and he had a title for Orders in 1537. Throughout his life he was a staunch Protestant and at Cambridge he used to attend the preachings of Hugh Latimer. We do not know how long Turner held his fellowship, possibly till his marriage with Jane daughter of George Ander, Alderman of Cambridge. He left Cambridge in 1540 and travelled about, preaching in various places. In Wood's *Athenae Oxonienses* we read, "In his rambles he settled for a time in Oxon among several of his countrymen that he found there, purposely for the conversation of men and books. . . . At the same time and after, following his old trade of preaching without a call, he was imprisoned for a considerable time."[1] On his release he left England and travelled in Italy, Germany and Holland. He tells us in his herbal that he visited Cremona, Como, Milan, Venice and Chiavenna, and at Bologna[2] he studied botany under Luca Ghini. Either there or at Ferrara he took his M.D. degree. From Italy he went to Zurich, where

[1] It has been suggested that Turner was imprisoned for his refusal to subscribe to the Six Articles and that he recanted to save his life. But, as Dr. B. D. Jackson has pointed out, Turner was made of sterner stuff and his whole life and writings are a standing contradiction to any such supposition.

[2] One of the earliest botanic gardens in Europe was at Bologna. It was founded by Luca Ghini. It is interesting to see how frequently Turner in his herbal quotes Ghini, and cites his authority against other commentators. Luca Ghini was the first who erected a separate professorial chair at Bologna for Botanical Science. He himself lectured on Dioscorides for twenty-eight years. He was the preceptor of Caesalpinus and Anquillara, two of the soundest critics on botanical writings of that age.

The most famous public botanical gardens in Europe during the sixteenth,

he formed his intimate and lifelong friendship with Conrad Gesner,[1] the famous Swiss naturalist.

He subsequently visited Basle and Cologne, and it was in these two cities that his small religious books upholding the Protestant cause were printed. They were very popular in England, so much so that in the last year of Henry VIII.'s reign they were prohibited. Turner spent some time botanising in the Rhine country: in his herbal he speaks of different plants which he collected at Bonn, Basel, Bingen, Cologne, "Erenffelde" and "Sieburg." Then he went to Holland and East Friesland—the latter he frequently mentions—and became physician to the "Erle of Emden." It was probably at this time that he explored the islands off the mainland. He was in correspondence with "Maister Riche and maister Morgan, Apotecaries of London," two names which, it is interesting to note, occur also in de l'Obel's works and in Gerard's Herbal.

Turner wrote the first part of his Herbal when he was

---

seventeenth and early eighteenth centuries were the following. I give them in the order in which they were made :—

1533—Padua.
1544—Florence.
1547—Bologna.
1570—Paris.
1598—Montpellier.
1628—Jena.
1632—Oxford.
1637—Upsala.
1673—Chelsea.
1675—Edinburgh.
1677—Leyden.
1682—Amsterdam.
1725—Utrecht.

The first botanic garden in America was founded in Philadelphia by John Bartram, the great American botanist, in the middle of the eighteenth century.

[1] Gesner had a high opinion of Turner, of whom he wrote :—

"Ante annos 15, aut circiter cum Anglicus ex Italia rediens me salutaret (Turnerus) is fuerit vir excellentis tum in re medica tum aliis plerisque disciplinis doctrinae aut alius quisquam vix satis memini."—*De Herbis Lunariis*, 1555.

abroad, but he delayed publication until the conclusion of his wanderings. On his return to England he became chaplain and physician to the Duke of Somerset, and it is generally believed that he sat in the House of Commons.[1] He was promised the prebend of Botevant in York, and in a letter written to thank Cecil for the promise we find the remark, "My chylder have bene fed so long with hope that they are very leane, i wold fayne have thē fatter if it were possible."

Turner held this appointment for little more than two years, and after failing to obtain either the provostship of Oriel College, Oxford, or the presidency of Magdalen, he seems to have become despondent. He wanted a house "where i may studie in and have sū place to lay my bookes in," and in another letter he complains of "being pened up in a chamber with all my ho[use] holde seruantes and children as shepe in a pyndfolde. . . . i can not go to my booke for ye crying of childer and noyse yt is made in my chamber." Finally he begged leave to go abroad, "where I will also finishe my great herball and my bookes of fishes, stones and metalles if God send me lyfe and helthe." He was subsequently made Dean of Wells, but he lost this office on the accession of Mary, and, like so many of the Protestant divines, he went abroad. He stayed at Bonn, Frankfort, Freiburg, Lauterburg [? Lauenburg], Mainz, Rodekirche, Strasburg, Speyer, Worms, Cologne and Weissenburg. At Cologne and Weissenburg he had gardens, and it was from Cologne that he published the second part of his Herbal. His works were proclaimed heretical for the second time in 1555, and the Wardens of every Company had to give notice of any copy they had in order that they might be destroyed. It is not surprising that Turner's works are rare!

On the accession of Elizabeth he returned to England and

[1] The Duke of Somerset was himself keenly interested in botanical investigations, and Turner frequently refers to the Duke's garden. It was during this time that Turner had his own garden at Kew. That he sat in the House of Commons is generally supposed from a passage in his *Spiritual Physik*, and this view is sustained by the character of the Hunter in his *Romish Wolfe*.

was reinstated in the deanery of Wells.[1] His diocesan seems to have found him troublesome, for in 1559 the Bishop of Bath and Wells wrote:

"I am much encombred with mr. Doctor Turner Deane of Welles for his indiscreete behavior in the pulpit where he medleth wth all matters. . . . I have advertised him by wrytynges and have admonished secretly by his owne frendes: notwithstanding he persisteth still in his follie: he conteñeth all Bishopps and calleth thē white coats, typpett gentlemē, with other wordes of reproche [mu]che more unsemlie and asketh 'who gave them autoritie more ouer me then I ouer them'?

"Gilbert Bath and Wells."

January 24,
1559–60.

There is a story told that Turner trained his dog at a given sign to snatch the bishop's square cap off his head when the prelate was dining with him. If this is true, possibly it accounts for the fact that he was subsequently suspended for Nonconformity, after which, being precluded from clerical duties, he left Wells and returned to London. He lived in Crutched Friars and, like the two other Elizabethan herbalists, had a famous garden. He was in failing health when he completed his herbal, and there is extant a pathetic letter (the greater part of it written by an amanuensis) to his staunch patron Lord Burleigh, which is signed "Your old and seikly client

wllm turner doctor of physic."

Turner died in 1568, and was buried in S. Olave's, Crutched Friars, where the tablet to his memory can still be seen.

[1] It has been asserted in some accounts of Turner that he was a Canon of Windsor, but this is a mistake. The Canon of Windsor was Richard Turner, also a Protestant, and, like the herbalist, exiled during Mary's reign.

CLARISSIMO . DOCTISSIMO . FORTISSIMOQUE . VIRO | GULIELMO .

TURNERO . MEDICO . AC . THEOLOGICO . PERITISSI | MO . DECANO . WELLENSI . PER . ANNOS . TRIGINTA . IN . VTRAQUE | SCIENTIA . EXERCITATISSIMVS . ECCLESIAE . ET . REI . PUBLICAE | PROFVIT . ET . CONTRA . VTRIVSQUE . PERNITIOSISSIMOS . HOS | TES . MAXIME . VERO . ROMANUM . ANTICHRISTVM . FORTISSIMUS | JESU . CHRISTI . MILES . ACERRIME . DIMICAVIT . AC . TANDEM . COR | PUS . SENIO . ET . LABORIBUS . CONFECTVM . IN . SPEM . BEATISSIM : |

RESVRRECTIONIS . HIC . DEPOSVIT . ANIMAM . IMMORTALEM | CHARISSIMO . EIVSQUE . SANCTISSIMO . DEO . REDDIDIT . ET . DEVICTIS . CHRISTI . VIRTUTE . MVNDI . CARNISQUE . VIRIBUS . TRIUMPHAT IN AETERNUM .

MAGNVS . APOLLINEA . QVONDAM . TVRNERVS . IN . ARTE
MAGNUS . ET . IN . VERA . RELIGIONE . FVIT .
MORS . TAMEN . OBREPENS . MAIOREM . REDDIDIT ILLVM .
CIVIS . ENIM . CAELI . REGNA . SVPERNA . TENET
OBIIT . 7 DIE . IVLII . AN . DOM . 1568 .

In his will, which is too long to quote here, Turner bequeathed to his wife [1] " his best pece or syluer vessell and halfe dozen of syluer spones," and to his nephew " my lyttell furred gowne." Peter, the son to whom he left " all my writen bookes and if he be a preacher all my diuinitie bookes, & yf he practise Phisicke all my physicke bookes," had some knowledge of plants, for in a copy of Turner's Herbal in the Linnean Society's Library there is a long list of errata for which Peter Turner apologised in an Address to the Reader. There is something very naïve and charming about Peter's admiration for his father's " fame and estimation." He tells us that he has diligently compared the printed book with his father's " owne hande copie," and refrains from having the whole book printed again because " I

[1] Turner's widow subsequently married Richard Cox, who became Bishop of Ely. She founded a scholarship at Cambridge in memory of her first husband.

should have done against Charitie to have caused the Printer by that meanes to lose all his labor and cost which he hath bestowed in printing hereof. Wherefore, gentle Reader, beare a little with the Printer that never was much accustomed to the printing of Englishe and afore thou reade over this booke correct it as I haue appointed and then the profit thereof will abundantly recompense thy paynes. In the meane time vse this Herbale in stede of a better and give all laude and prayse unto the Lorde."

Turner was the first Englishman who studied plants scientifically, and his herbal marks the beginning of the science of botany in England. Like most writers of any value, he impressed his personality upon his books, and these show him to have been a man of indomitable character, caustic wit and independent thought. "Vir solidae eruditionis judicii" he is called by John Ray. His first botanical work was the *Libellus de re herbaria novus* (1538), printed by John Byddell in London. This little book is particularly interesting, because it is the first in which localities of native British plants are given. In 1548 he published another small book entitled *The names of herbes in Greke, Latin, Englishe, Duche, and Frenche wyth the commone names that Herbaries and Apotecaries use, gathered by William Turner.* In the preface he tells us that he had begun to "set furth an herbal in latyn," but that when he asked the advice of physicians, "their advise was that I shoulde cease from settynge out of this boke in latin till I had sene those places of Englande, wherein is moste plentie of herbes, that I might in my herbal declare to the greate honoure of our countre what numbre of sovereine and strang herbes were in England, that were not in other nations, whose counsell I have folowed, deferrying to set out my herbal in latyn, tyl that I have sene the west countrey, which I never sawe yet in al my lyfe, which countrey of al places of England, as I heare say, is moste richely replenished wyth al kindes of straunge and wonderfull workes and giftes of nature as are stones, herbes, fishes and metalles, when as they that moued me to the settyng furth of my latin herbal, hearde this so reasonable an excuse, they moved me to set out an herbal in Englishe as Fuchsius dyd in latine wyth the discriptions, figures and properties of as many herbes, as I had sene and knewe, to whom I could make no other answere but that I had no such leasure in this vocation and place that I am nowe in, as is neccessary for a man that shoulde take in hande suche an interprise. But thys excuse coulde not be admitted for both certeine scholars, poticaries, and also surgeons, required of me if that I woulde not set furth my latin herbal, before I have sene the west partes, and have no leasure in thys place and vocation to write so great a worke, at the least to set furth my judgement of the names of so many herbes as I knew, whose request I have accomplished, and have made a litle boke, which is no more but a table or regestre of suche bokes as I intende by the grace of God to set furth hereafter; if that I may obteine by your graces healp such libertie and leasure with convenient place, as shall be necessary for suche a purpose."

Turner's notable work, his Herbal, is the only original work on botany written by any Englishman in the sixteenth century. The first part of it was printed in London by Steven Mierdman, a Protestant refugee from Antwerp, in 1551. The second part was printed by Arnold Birckman, at Cologne, in 1561, during Turner's enforced exile. Birckman also printed the edition of 1568, which contained all three parts. (For the full title, etc., see Bibliography of Herbals, p. 208.)

One of the most attractive features of this Herbal is the number of beautiful woodcuts with which it is illustrated. A few were specially drawn and cut for the author, but the great majority are reproductions of the exquisite drawings in Fuchs's herbals (*De historia Stirpium*, 1545; and *Neue Kreüterbuch*, 1543). Nearly all the illustrations in the famous sixteenth-century Flemish, English and Swiss herbals were printed from the actual wood-blocks or copied from the illustrations in Fuchs's works. Notably in Hieronymus Bock's *Kreüter Buch* (1546), Rembert Dodoens's *Cruÿdtboeck* (1554), Henry Lyte's *Niewe Herball* (1578), and Jean Bauhin's *Historia plantarum universalis* (1651). It is a remarkable fact that so far as wood-engraving is concerned this country has contributed nothing to the art of plant illustration. In the first English illustrated Herbal, the *Grete Herball* of 1526, the figures are merely copies of the inferior cuts in the later editions of the *Herbarius zu Teutsch*, and, with the exception of Parkinson's *Paradisus*, all the English sixteenth- and early seventeenth-century herbals borrowed their illustrations from Flemish or German sources. Fuchs had two sets of blocks for his Herbal, one for the folio edition of 1542 and the other for the octavo edition of 1545. It was the blocks for the latter which were borrowed by Turner's printer, and it has been suggested that it was his desire to secure these beautiful illustrations which led him to have his herbal printed at Cologne.[1] Over 400 of Fuchs's blocks were used in the complete edition of Turner's Herbal, and, of the rest, some are copied from the smaller figures in Mattioli's [2] commentary on Dioscorides.

Turner dedicated the first part of his Herbal (1551) to the Duke of Somerset, uncle to Edward VI., and at that time Lord Protector. The preface is delightful and I quote a part of it :—

"To the mighty and christiane Prince Edward, Duke of Summerset, Erle of Herford, Lorde Beauchampe, and Uncle unto the Kynges maiesty, Wyllyam Turner his servant wysheth increase in the knowledge of Goddes holy worde and grace to lyue thereafter. Although (most myghty and Christian Prince) there be many noble and excellent actes and sciences, which no

[1] It was for the same reason that Henry Lyte's translation of Dodoens was printed abroad.

[2] Pierandrea Mattioli (1501–1577) was physician successively to the Archduke Ferdinand and to the Emperor Maximilian II. With the exception of Fabio Colonna he was the greatest of the Italian herbalists.

man douteth but that almyghty God, the author of all goodness, hath gyuen unto us by the hands of the Hethen, as necessary unto the use of Mankynd: yet is there none among them all, whych is so openly cōmended by the verdit of any holy writer in the Bible, as is ye knowledge of plantes, herbes, and trees and of Phisick. I do not remembre that I have red anye expressed commendations of Grammer, Logick, Philosophie, naturall or morall, Astronomie, Arithmetyke, Geometry, Cosmographie, Musycke, Perspectiue or any other such lyke science. But I rede amonge the commendatyons and prayses of Kyng Salomon, that he was sene in herbes shrubbes and trees and so perfectly that he disputed wysely of them from the hyghest to the lowest, that is from the Cedre tre in Mount Liban unto the Hysop that groweth furth of the wall. If the Knowledge of Herbes, shrubbes and trees which is not the lest necessary thynge unto the knowledge of Phisicke were not greatly commendable it shulde never have bene set among Salomon's commendacyons and amongst the singular giftes of God. Therefor whereas Salomon was commended for the Knowledge of Herbes the same Knowledge was expressedly ynough cōmended there also." Continuing, he speaks of learned Englishmen "Doctor Clement, Doctor Wendy and Doctor Owen, Doctor Wolton and Maister Falconer"[1] which "have as much knowledge in herbes yea and more than diuerse Italianes and Germanes whyche have set forth in prynte Herballes and bokes of simples. Yet hath none of al these set furth any thyng other to the generall profit of hole Christendome in latin and to the honor of thys realme, nether in Englysh to the proper profit of their naturall countre." After slyly observing that perhaps they do not care to jeopardise their estimation, he compares himself, for having ventured to write this book, with the soldier "who is more frendly unto the commonwealth, which adventurously runneth among the

[1] This was probably the John Falconer who sent English plants to Amatus Lusitanus, who taught physic at Ferrara and Ancona, and whose commentary on Dioscorides was published in 1553.

myddes of hys enemyes, both gyuyng and takyng blowes, then he that, whilse other men feight, standeth in the top of a tre iudging how other men do, he beynge without the danger of gonne shot himself."

To those who may object that it is too small, he explains that he will write more fully when he has "travelled diverse shyres in England to learn more of the herbs that grow there." Others may condemn him for writing in English, "for now (say they) every man without any study of necessary artes unto the knowledge of Phisick will become a Phisician . . . euery man nay euery old wyfe will presume, not without the mordre of many, to practyse Phisick." To these he succinctly replies, "How many surgianes and apothecaries are there in England which can understand Plini in Latin or Galen and Dioscorides?" The English physicians, he says, rely on the apothecaries, and they in turn on the old wives who gather the herbs. Moreover, since the physicians are not present when their prescriptions are made up, "many a good mā by ignorance is put in jeopardy of his life, or good medecine is marred to the great dishonesty both of the Phisician and of Goddes worthy creatures." All this can be avoided by having a herbal written in English. Dioscorides and Galen, he points out, wrote in their native tongue, Greek. "Dyd Dioscorides and Galen give occasion for every old wyfe to take in hād the practise of Phisick? Did they giue any iust occasion of murther? If they gaue no occasyon unto every old wyfe to practise physike then give I none. If they gave no occasion of murther then gyue I none . . . then am I no hynderer wryting unto the English my countremen an English herball."

The second part of Turner's Herbal is dedicated to his old patron, Thomas Lord Wentworth, and the complete work, including the third part, to Queen Elizabeth.[1] In the preface

[1] Queen Elizabeth's love of gardening and her botanical knowledge were celebrated in a long Latin poem by an Italian who visited England in 1586 and wrote under the name of Melissus (see *Archæologia*, VII. 120).

to this last he reminds the queen of a conversation he had had with her in Latin eighteen years before, at the Duke of Somerset's house, when he was physician to that nobleman. It is in this preface also that he criticises the foreign herbalists; though he has learnt much from them, they had much to learn from him, " as their second editions maye testifye." He claims that in the first part of his herbal he taught " the truth of certeyne plants which these above-named writers (Matthiolus, Fuchsius, Tragus and Dodoneus) either knew not at al or ellis erred in them greatlye. . . . And because I would not be lyke unto a cryer yt cryeth a loste horse in the marketh, and telleth all the markes and tokens that he hath, and yet never sawe the horse, nether coulde knowe the horse if he sawe him : I wente into Italye and into diverse partes of Germany, to knowe and se the herbes my selfe."

The book owes much of its charm to its vivid descriptions of the plants, and the fascinating and unexpected details he gives us about them. The comparison of dodder, for instance, to " a great red harpe strynge," is a happy touch which it is impossible to forget. " Doder groweth out of herbes and small bushes as miscelto groweth out of trees. Doder is lyke a great red harpe strynge and it wyndeth about herbes foldyng mych about them and hath floures and knoppes one from an other a good space. . . . The herbes that I have marked doder to growe most in are flax and tares."

These accurate observations and careful descriptions are characteristic of the writer, and recall similar touches in the Saxon herbals. For example, he records that the stamens of the Madonna lily have a different smell from the flower itself, and that the berries of the bay tree are almost, but not quite, round. There is only space to quote the following :—

" The lily hath a long stalk and seldom more than one, howbeit it hath somtyme II. It is II or III cubites hyghe. It hath long leves and somthyng of the fashion of the great satyrion. The flour is excedyng white and it hath the forme or fashion of a long quiver, that is to say, smal at the one end and byg at the other. The leves of the floures are full of crestes, and the overmost ends of the leves bowe a little backwarde and from the lowest parte within come forth long small yelow thynges lyke thredes of another smelle than the floures are of. The roote is round and one pece groweth hard to another allmoste after the maner of the roote of Garleke, but that the clowes in the lily are broder."

"The leaves of the Bay tree are alwayes grene and in figure and fashion they are lyke unto periwincle. They are long and brodest in the middest of the lefe. They are blackishe grene namely when they are olde. They are curled about the edges, they smell well. And when they are casten into the fyre they crake wonderfully. The tre in England is no great tre, but it thryveth there many partes better and is lustier than in Germany. The berries are allmoste round but not altogether. The kirnell is covered with a thick black barke which may well be parted from the kirnell."

"Blewbottel groweth in ye corne, it hath a stalke full of corners, a narrow and long leefe. In the top of the stalke is a knoppy head whereupon growe bleweflowers about midsummer. The chylder use to make garlandes of the floure. It groweth much amonge Rye wherefore I thinke that good ry in an evell and unseasonable yere doth go out of kinde in to this wede."

"Pennyroyal.—It crepeth much upon the ground and hath many lytle round leves not unlyke unto the leves of merierum gentil but that they are a little longer and sharper and also litle indented rounde about, and grener than the leves of meriurum ar. The leves grow in litle branches even from the roote of certayn ioyntes by equall spaces one devyded from an other. Where as the leves grow in litle tuftes upon the over partes of the braunches. . . . Pennyroyal groweth much, without any setting, besyd hundsley [Hounslow] upon the heth beside a watery place."

Of camomile he writes: "It hath floures wonderfully shynynge yellow and resemblynge the appell of an eye . . . the herbe may be called in English, golden floure. It will restore a man to hys color shortly yf a man after the longe use of the bathe drynke of it after he is come forthe oute of the bath. This herbe is scarce in Germany but in England it is so plenteous that it groweth not only in gardynes but also VIII mile above London, it groweth in the wylde felde, in Rychmonde grene, in Brantfurde grene. . . . Thys herb was consecrated by the wyse men of Egypt unto the sonne and was rekened to be the only remedy of all agues."

Unlike modern authorities, Turner contends that our English hyssop is the same plant as that mentioned in the Bible, and he also describes a species which does not now exist. "We have in Sumershire beside ye cōmē Hysop that groweth in all other places of Englande a kinde of Hysop that is al roughe and hory and it is greater muche and stronger then the cōmen Hysop is, som call it rough Hysop." Another plant which seems to have disappeared and which, he states, no other writer describes, is "the wonderful great cole with leaves thrise as thike as ever I saw any other cole have. It hath whyte floures and round berryes lyke yvy. This herbe groweth at douer harde by the Sea-syde. I name it the Douer cole because I founde it first besyde Douer." Incidentally he mentions samphire also as growing at Dover.

It is interesting to find that Turner identifies the *Herba Britannica* of Dioscorides and Pliny (famed for having cured the soldiers of Julius Cæsar of scurvy in the Rhine country) with *Polygonum bistorta*, which he observed plentifully in Friesland, the scene of Pliny's observations. This herb is held by more modern authorities to be *Rumex aquaticus* (great water dock).

Throughout the Herbal there are recollections of the north of England, where the author spent his boyhood. Of heath, for instance, he tells us: "The hyest hethe that ever I saw

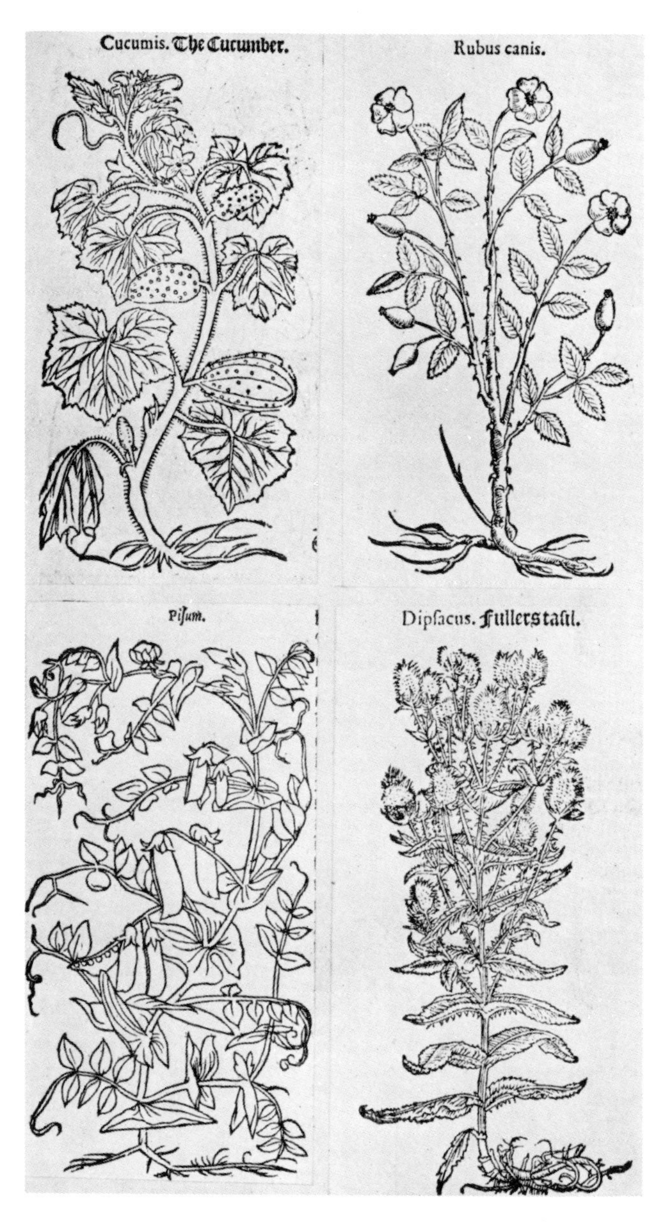

ILLUSTRATIONS FROM TURNER'S "HERBAL"

growweth in Northumberland, which is so hyghe that a man may hyde himself in." Of the wild hyacinth he writes: "The boyes in Northūberland scrape the roote of the herbe and glew theyr arrowes and bokes wyth that slyme that they scrape of." Of sea-wrake (seaweed) he tells us: "In the Bishopriche of Durham the housbandmen of the countie that dwel by the sea syde use to fate [fatten, *i. e.* manure] their lande with seawrake." Under "birch" we find: "Fisherers in Northumberland pyll off the uttermost barke and put it in the clyft of a sticke and set in fyre and hold it at the water syde and make fish come thether, whiche if they se they stryke with theyr leysters or sammon speres. The same," he continues, "is good to make hoopes of and twigges for baskettes, it is so bowinge. It serveth for many good uses and for none better then for betinge of stubborne boyes that ether lye or will not learne."

Cudweed "is called in Northumberland chafwede because it is thought to be good for chafyng of any man's fleshe wyth goynge or rydynge." And it would be interesting to know if the daisy is still called banwurt in the north. "The Northern men call thys herbe banwurt because it helpeth bones to knyt agayne. . . . Plinie writeth that the dasey hath III and sometimes IV little whyte leves whiche go about the yelow knope, it appereth that the double Daseys were not founde in plinies tyme whych have a greate dele mo then Plini maketh mention of."

There are other country customs which he records. "Shepherds use clivers [goosegrass] in stede of a strayner to pull out here of the mylke;" "birderers [bird-catchers] take bowes of birch and lime the twigges and go a bat folinge with them;" "som make a lee [lye] or an ashy water of the rotes of gentian wherwyth they toke out spottes very well out of cloth." He mentions woad as "trimmed wyth mannes labor in dyenge and wull and clothe," and teazle "which the fullers dresse their cloth wtall." Apparently Turner gave the spindle tree its name, for he says: "I coulde never learne an Englishe name for it. The Duche men call it in Netherlande, spilboome, that is, spindel tree, because they use to make spindels of it in that countrey, and me thynke it maye be so well named in English seying we have no other name. . . . I know no good propertie that this tree hath, saving only it is good to make spindels and brid of cages " [bird cages].

The use of complexion washes was a custom on which Turner was alarmingly severe. There are fewer beauty recipes in his herbal than in any other—only four altogether. " Some weomen," we find, "sprinkle ye floures of cowslip w^t^ whyte wine and after still it and wash their faces w^t^ that water to drive wrinkles away and to make them fayre in the eyes of the worlde rather then in the eyes of God, whom they are not afrayd to offend." And of marygold we learn that " Summe use to make theyr here yelow with the floure of this herbe, not beyng contēt with the naturall colour which God hath geven thē."

There is curiously little folk lore in this herbal, and most of it is guarded by " some do say " or " some hold." Nevertheless, with this qualification, Turner gives us fragments of folk lore not to be found in other herbals. For instance, that nutshells burnt and bound to the back of a child's head will make grey eyes black, and that parsley thrown into fish ponds will heal the sick fishes therein. Again, this is the first herbal in which any account is to be found of the very old custom of curing disease in cattle by boring a hole in the ear and inserting the herb bearfoot.[1]

" They say it should be used thus. The brodest part of the ear must have a round circle made about it w^t^ the blood that rinneth furth with a brasen botken and the same circle must be round lyke unto the letter O, and when this is done without and in the higher part of the ear the halfe of the foresaid circle is to be bored thorowe with the foresaid botken and the roote of the herbe is to be put in at the hole, when y^t^ newe wounde that hath receyued it holdeth it so fast, that it will

[1] Parkinson in his *Theatrum Botanicum* also mentions this use of bearfoot.

not let it go furth, then all the mighte and pestilent poison of the disease is brought so into the eare. And whilse the part which is circled aboute dyeth and falleth awaye y[t] hole beast is saved with the lose of a very small parte."

Another piece of folk lore is remarkable because it is the only instance in an English herbal of a belief in the effect of a human being on a plant: "If ye woulde fayne have very large and greate gourdes, then take sedes that growe there [in the sides]. . . . And let weomen nether touche the yonge gourdes nor loke upon them, for the only touchinge and sighte of weomen kille the yonge gourdes." This belief he quotes from Pliny.

Turner, again, is the only old herbalist who refers to the old and widespread belief that larch was fire-proof. It was largely used, he tells us, for laying under the tiles of newly-built houses, as "a sure defence against burning," and he narrates at length how Julius Cæsar was unable to burn a tower built with larch. On the old mandrake legend he is scathing. "The rootes which are counterfited and made like litle puppettes and mammettes which come to be sold in England in boxes with heir [hair] and such forme as man hath, are nothyng elles but folishe fened trifles and not naturall. For they are so trymmed of crafty theves to mocke the poore people withall and to rob them both of theyr wit and theyr money. I have in my tyme at diverse tymes takē up the rootes of mandrake out of the grounde but I never saw any such thyng upon or in them as are in and upon the pedlers rootes that are comenly to be solde in boxes. It groweth not under galloses [gallows] as a certayn doting doctor of Colon in his physick lecture dyd teach hys auditors." But he accepts without question the belief in its efficacy as an anæsthetic: "It is given to those who must be burned or cut in some place that they should not fele the burning or cuttyng." Of wine made of it, he says: "If they drynk thys drynke they shall fele no payne, but they shall fall into a forgetfull and slepishe drowsiness. Of the apples of mandrake, if a man smell of them thei will make hym slepe and also if they be eaten.

But they that smell to muche of the apples become dum . . . thys herbe diverse wayes taken is very jepardus for a man and may kill hym if he eat it or drynk it out of measure and have no remedy from it. . . . If mandragora be taken out of measure by and by slepe ensueth and a great lousing of the streyngthe with a forgetfulness."

Turner is one of the few herbalists who cautions against the excessive use of any herb. "Onions eaten in meat largely make the head ake, they make them forgetfull whiche in the tyme of syknes use them out of mesure." "Cole engendreth euell and melancholie juice. It dulleth the syght and it troubleth the slepe wyth contrary thynges which are sene in the dreme." Of nigella he writes: "Take hede that ye take not to muche of this herbe, for if ye go beyonde the mesure it bryngeth deth." "Hemp seed," he says, "if it be taken out of measure taketh men's wyttes from thē as coriander doth." "If any person use saffron measurably it maketh in them a good colour, but if thei use it out of mesure it maketh hym loke pale, and maketh the hede ache and hurteth the appetite." For those who have taken an overdose of opium there is a surprising remedy. "If the pacient be to much slepi put stynkynge thynges unto hys nose to waken hym therewith." As in all herbals of this period, there are an astonishing number of remedies against melancholy and suggestions for those whose weak brains will not stand much strong drink; but, while remedies for broken heads, so common in the older herbals, are conspicuously absent, we find that walnuts are recommended "for the bytings both of men and dogges"!

As in the *Grete Herball,* there are many descriptions of other substances besides herbs, some of the longest being of dates, rice, olives, citron, pomegranates and lentils. The account of citron it would be pleasant to transcribe in full, not for the sake of the story but for the manner of the telling. One could listen to a sermon of considerable length from a divine who, in a book intended for grown-ups, has a tale of "two naughty murthering robbers, condemned for theyr murder and robery to be flayn and poysoned to deth of serpentes, and such venemous bestes," and of the one who, owing to having eaten "a pece of citron," remained, Daniel-like, unhurt by the poison of the snakes, whilst the other who had not taken this precaution "fell down sterk dede." And finally, the moral—"Wherefore it were wisdome that noblemen and other that are bydden to dynner of theyr enemies or suspected frendes before they eat any other thyng should take a pece of citron."

The later sixteenth-century herbalists owed much to the famous herbalists of the Netherlands, and above all to that prince amongst publishers, Christophe Plantin of Antwerp, whose personality secured him a unique place in the literary world. Indeed, there is a splendour about the works of the Flemish herbalists unequalled by any others of this period, with the exception of the Bavarian doctor Leonhard Fuchs. There is no comparison between them and the Italian herbalists of the Renaissance, who, for the greater part, devoted themselves to studying the classical writers and identifying the plants mentioned by the old authorities. France, curiously enough, contributed comparatively little when the herbal was at its zenith, though it must of course be remembered that the Bauhins, who rank as Swiss herbalists, were of French extraction. But it is difficult to estimate the influence of the works of those three notable friends, Rembert Dodoens, Charles de l'Escluse and Matthias de l'Obel, particularly on the English herbalists. The most famous English herbal—Gerard's—is virtually a translation of the *Pemptades* of Dodoens. Lyte's translation of the *Cruydtboeck* was the standard work on herbs during the latter part of the century, and Parkinson incorporated a large part of de l'Obel's unfinished book in his *Theatrum Botanicum*.

De l'Obel, after whom the little garden flower—lobelia—is named, spent the greater part of his life in England. He was a Fleming by birth and a doctor by profession,[1] and he was

[1] He studied medicine at Montpelier under Guillaume Rondelet, who bequeathed him his botanical manuscripts. D'Aléchamps, Pena and Jean Bauhin, all famous herbalists, were also pupils of Rondelet.

physician to William the Silent until his assassination. About 1569 he came over to England (with his friend Pena, who at one time was physician to Louis XIII.) and lived at Highgate with his son-in-law. He superintended Lord Zouche's garden at Hackney, and later was given the title of botanist to James I. L'Obel's great work, written in collaboration with Pena, was the *Stirpium Adversaria Nova*, printed in London by Thomas Purfoot in 1571.[1] Pulteney, in his *Biographical Sketches* (1790), makes the extraordinary statement that Christophe Plantin of Antwerp was the real printer. It has, however, been pointed out by modern authorities that the archives of the Plantin Museum show that Plantin bought 800 copies of Purfoot's edition, with the wood blocks, for 1320 florins. In 1576 Plantin published de l'Obel's *Plantarum seu Stirpium historia*, and to this he appended the first part of the *Adversaria*, keeping Purfoot's original colophon.

Although Dodoens neither lived in England nor had any of his works printed here, his *Cruÿdtboeck* became one of the standard works in this country through Lyte's translation. Dodoens was born at Malines about 1517 and, after studying at Louvain, visited the universities and medical schools of France, Italy and Germany, graduated M.D., and was appointed physician to Maximilian II. and Rudolf II. successively. In the latter part of his life he was Professor of Medicine at Leyden, where he died in 1585. Plantin published Dodoens's most important work, *Stirpium historiae pemptades sex sive libri triginta*, in which some of the figures are copied from the fifth-century manuscript[2] copy of Dioscorides. Dodoens's first book, the *Cruÿdtboeck*, was translated into French by his friend Charles de l'Escluse[3] and afterwards into English by Henry Lyte.

[1] For full title see Bibliography of Herbals, p. 210.

[2] This manuscript, now in the Vienna Library, was bought from a Jew in Constantinople for 100 ducats by Auger-Geslain Busbecq, when he was on a mission to Turkey.

[3] On one of his visits to England de l'Escluse met Sir Francis Drake, who gave him plants from the New World.

Lyte, who was an Oxford man, travelled extensively in his youth and made a collection of rare plants. He contributed nothing original to the literature on herbs, but his translation of the French version of the *Cruydtboeck* was an inestimable service. His own copy of the French version, which is now in the British Museum, has on the title-page the quaint inscription "Henry Lyte taught me to speake Englishe." The book is full of MS. notes and references to Turner.

The full title of Lyte's book is as follows: "A niewe Herball or Historie of Plantes: wherein is contayned the whole discourse and perfect description of all sortes of Herbes and Plantes: their divers and sundry kindes: their straunge Figures, Fashions and Shapes: their Names, Natures, Operations, and Vertues: and that not only of those which are here growyng in this our Countrie of Englande but of all others also of forrayne Realmes, commonly used in Physicke. First set foorth in the Doutche or Almaigne tongue by that learned D. Rembert Dodoens, Physition to the Emperour: And nowe first translated out of French into English by Henry Lyte Esquyer."

(*Colophon.*) "Imprinted at Antwerpe by Me Henry Loë Booke printer and are to be solde at London in Paul's churchyarde by Gerard Dewes."[1]

The beautiful illustrations in Lyte's *Dodoens* are to a large extent printed from the same blocks as those in the octavo edition (1545) of Fuchs. In Fuchs there are about 516 illustrations, and in Lyte's *Dodoens* about 870. Those which are not copied from Fuchs were probably collected by Dodoens himself, who, according to some verses at the beginning of the herbal, took a practical interest in the publication of the English translation of his book.

> "Till Rembert he did sende additions store,
> For to augment Lyte's travell past before."

The original wood-blocks never came to England, and three

[1] For subsequent editions see Bibliography of Herbals, p. 211.

years later van der Loë's widow sold them to Christophe Plantin for 420 florins.

All the commendatory verses at the beginning of Lyte's herbal are in Latin, except some lines in which William Clowes speaks of writing about herbs as " a fit occupation for gentlemen and wights of worthy fame," and recalls the great men who have immortalised themselves thereby, notably Gentius, Lysimachus, Mythridates and Dioscorides. Then, after giving due praise to Dodoens, " Whose learned skill hath offered first this worthy worke to vewe," Clowes ends with four lines in which he plays upon the name of the translator :

> " And Lyte, whose toyle hath not bene light to dye it in this grayne,
> Deserves no light regarde of us : but thankes and thankes agayne.
> And sure I am all English hartes that lyke of Physickes lore
> Will also lyke this gentleman : and thanke hym muche therefore."

The herbal is dedicated to Queen Elizabeth " as the best token of love and diligence that I am at this time able to shew. . . . And doubtless if my skill in the translation were answerable to the worthynesse eyther of the Historie itselfe or of the Authours thereof I doubt not but I should be thought to haue honoured your Maiestie with an acceptable present." The preface is dated from " my poore house at Lytes carie within your Maiesties Countie of Somerset the first day of Januarie MDLXXVIII."

In 1606 there appeared the book commonly known as *Ram's little Dodoen*. It purported to be an epitome of Lyte's *Dodoens*, but, though some of its matter has been abridged from Dodoens's work, it is in reality a compilation of recipes unworthy of the great name it bears. In his preface the author tells us : " I have bestowed some tyme in reducing the most exquisit new herball or history of plants (first set forth in Dutch and Almayne tongue by the learned and worthy man of famous memory Dr. Rembert Dodeon (*sic*) Phisician to the Emperor, and translated into English by Master Henry Lyte Esq.), with a brief and short epitome . . . so as where the great booke at large is not to be had but at a great price, which cannot be procured by the poorer sort, my endevor herein hath bin chiefly to make the benefit of so good, necessary and profitable a worke to be brought within the reach and compasse as well of you my poore countrymen and women whose lives, healths, ease and welfare is to be regarded with the rest, at a smaller price than the great volume is. My onely and greatest care hath byn of long tyme to knowe or thinke how and upon whome to bestow the dedication of this my small labour. And in the penning of this my letter my Affections are satisfied with the dedication thereof to these my poore and loving countrymen whosoever and to whose hands soever it may come. For whose sake I have desired publicatiō of the same, beseeching Almighty God to blesse us all."

The book is curiously arranged, for on one page we have "the practice of Dodoen," and on the opposite "the practises of others for the same Phisike helpes, collected and presented to the Author of this Treatise." There are directions for each month, and each is headed by a motto. The twelve mottoes, when read together, form the following quaint rhyme:—

> "January. With this fyre I warme my hand
> February. With this spade I digge my land
> March. Here I cut my Vine spring
> April. Here I hear the birds sing
> May. I am as fresh as bird on bough
> June. Corn is weeded well enough
> July. With this sithe my grasse I mowe
> August. Here I cut my corne full lowe
> September. With this flaile I earne my bread
> October. Here I sowe my wheats so red
> November. With this axe I kill my swine
> December. And here I brew both ale and wine."

There are some things in this little handbook worthy of remembrance, notably an imaginative passage in which the author tells us that "herbs that grow in the fields are better than those which grow in gardens, and of those herbs which grow in the fieldes, such as grow on hilles are best."

CHAPTER IV

GERARD'S HERBAL

" If odours may worke satisfaction, they are so soveraigne in plants and so comfortable that no confection of the apothecaries can equall their excellent vertue."—*Gerard's Herbal*, 1597.

WHEN one looks at the dingy, if picturesque, thoroughfare of Fetter Lane it is difficult to realise that it was once the site of Gerard's garden, and it is pleasant to remember that the city of London in those far-off days was as noted for the beauty of its gardens as for its stately houses. The owner of this particular garden in Fetter Lane is the most famous of all the English herbalists. His Herbal,[1] which was published in 1597, gripped the imagination of the English garden-loving world, and now, after the lapse of three hundred years, it still retains its hold on us. There are English-speaking people the world over who may know nothing of any other, but at least by name they know Gerard's Herbal. In spite of the condemnation he has justly earned, not only in modern times, but from the critics of his own day, for having used Dr. Priest's translation of Dodoens's *Pemptades* without acknowledgment, no one can wander in the mazes of Gerard's monumental book without succumbing to its fascination. One reads his critics with the respect due to their superior learning, and then returns to Gerard's Herbal with the comfortable sensation of slipping away from a boring sermon into the pleasant spaciousness of an old-fashioned fairy-tale. For the majority of us are not scientific, nor do we care very

[1] Americans who have the proud distinction of being " of Royal Indian descent " may be interested to know that a copy of Gerard's Herbal in Oxford has been identified as having belonged to Dorothy Rolfe, the mother-in-law of the Princess Pocahontas.

much about being instructed. What we like is to read about daffodils and violets and gilliflowers and rosemary and thyme and all the other delicious old-fashioned English flowers. And when we can read about them in the matchless Elizabethan English we ask nothing more. Who that has read it once can forget those words in the preface?—

"What greater delight is there than to behold the earth apparelled with plants as with a robe of embroidered works, set with Orient pearls and garnished with great diversitie of rare and costly jewels? But these delights are in the outward senses. The principal delight is in the minde, singularly enriched with the knowledge of these visible things, setting forth to us the invisible wisdome and admirable workmanship of almighty God."

And could any modern writer give with such simplicity and charm the "atmosphere" of the violet?

"Addressing myself unto the violets called the blacke or purple violets or March violete of the Garden, which have a great prerogative above others, not only because the minde conceiveth a certaine pleasure and recreation by smelling and handling of these most odoriferous flowers, but also that very many by these violets receive ornament and comely Grace: for there be made of them garlands for the head, nosegaies and posies, which are delightful to look on and pleasant to smell, speaking nothing of the appropriate vertues; yea Gardens themselves receive by these the greatest ornament of all, chiefest beautie and most gallant grace; and the recreation of the Minde which is taken heereby, cannot bee but verie good and honest; for flowers through their beautie, varietie of colour and exquisite formes do bring to a liberall and gentlemanly minde the remembrance of honestie, comeliness and all kindes of vertues. For it would be an unseemly thing, as a certain wise man saith, for him that doth looke upon and handle faire and beautifull things, and who frequenteth and is conversant in faire and beautifull places to have his minde not faire."

The bones, so to speak, of Gerard's work are, it is true, taken from Dodoens's splendid Latin herbal, but it is Gerard's own additions which have given the book its hold on our affections. He describes with such simplicity and charm the localities where various plants are to be found, and he gives so much contemporary folk lore that before we have been reading long we feel as though we were wandering about in Elizabethan England with a wholly delightful companion.

We know from Gerard's coat of arms that he was descended from a younger branch of the Gerards of Ince, a Lancashire family, but there are no records at the College of Arms to show his parentage. His name is frequently spelt with an e at the end, but Gerard himself and his friends invariably spelt it without. (The spelling "Gerarde" on the title-page of the Herbal is probably an engraver's error.) John Gerard was born at Nantwich in Cheshire in 1545, and educated at the school at Wisterson or Willaston, two miles from his native town. In the Herbal he gives us two glimpses of his boyhood. Under raspberry we find :—

"Raspis groweth not wilde that I know of. . . . I found it among the bushes of a causey neere unto a village called Wisterson, where I went to schoole, two miles from the Nantwich in Cheshire."

Writing of yew [1] he tells us :—

"They say that if any doe sleepe under the shadow thereof it causeth sickness and sometimes death and that if birds do eat of the fruit thereof it causeth them to cast their feathers and many times to die. All which I dare boldly affirme is altogether untrue : for when I was young and went to schoole divers of

[1] Yew berries are an ingredient in at least one prescription in a Saxon herbal (*Leech Book of Bald,* I. 63).

my schoole-fellowes and likewise myselfe did eat our fils of the berries of this tree and have not only slept under the shadow thereof but among the branches also, without any hurt at all, and that not one time but many times."

It is supposed that at an early age he studied medicine. In his Herbal he speaks of having travelled to Moscow, Denmark, Sweden and Poland, and it is possible that he went abroad as a ship's surgeon. This, however, is mere surmise. We know that in 1562 he was apprenticed to Alexander Mason, who evidently had a large practice, for he was twice warden of the Barber-Surgeons' Company. Gerard was admitted to the freedom of the same company in 1569.[1] Before 1577 he must have settled in London, for in his Herbal he tells us that for twenty years he had superintended the gardens belonging to Lord Burleigh in the Strand and at Theobalds in Hertfordshire. Hentzner, in his *Itinerarium*, gives a lengthy account of the gardens at Theobalds when Gerard was superintendent.

Gerard's own house was in Holborn and, as already mentioned, his garden, where he had over a thousand different herbs, was in what is now Fetter Lane.[2] What a wonderful garden that must have been, and how full it was of " rarities," ranging from white thyme to the double-flowered peach. How often we read of various plants, " these be strangers in England yet I have them in my garden," sometimes with the triumphant addition, " where they flourish as in their natural place of growing." In

[1] Gerard endeavoured to induce the Barber-Surgeons' Company to establish a garden for the cultivation and study of medicinal plants, but nothing came of the scheme.

[2] Formerly it was generally supposed that Gerard's garden was on the northern side of Holborn, but this is unlikely, for during the latter part of Elizabeth's reign the part which is now known as Ely Place and Hatton Garden was an estate of forty acres belonging to the Bishopric of Ely. Holborn was almost a village then, and Gerard tells us in his *Herball* that in Gray's Inn Lane he gathered mallow, shepherd's purse, sweet woodruff, bugle and Paul's betony, and in the meadows near red-flowered clary, white saxifrage, the sad-coloured rocket, yarrow, lesser hawkweed and the curious strawberry-headed trefoil. Wallflower and golden stonecrop grew on the houses.

1596 Gerard published a catalogue of twenty-four pages of the plants in this garden—the first complete catalogue of the plants in any garden, public or private.[1] A second edition was published in 1599. Of Gerard's knowledge of plants the members of his own profession had a high opinion. George Baker, one of the "chief chirurgions in ordinarie" to Queen Elizabeth, wrote of him: "I protest upon my conscience that I do not thinke for the Knowledge of plants that he is inferior to any, for I did once see him tried with one of the best strangers that ever came into England and was accounted in Paris the onely man,[2] being recommended to me by that famous man M. Amb. Parens; and he being here was desirous to go abroad with some of our herbarists, for the whiche I was the means to bring them together, and one whole day we spent therein, searching the most rarest simples: but when it came to the triall my French man did not know one to his fower." In 1598, the year after the publication of his Herbal, and again in 1607, Gerard was appointed examiner of candidates for admission to the freedom of the Barber-Surgeons' Company, but apart from this we have little definite knowledge of his life. He seems to have been a well-known figure in the later years of Elizabeth and the early years of James I., and it is probable that he held the same position in the household of Robert Earl of Salisbury, Secretary of State, as he had held in that of his father, Lord Burleigh. A few years before he died James's queen (Anne of Denmark) granted him the lease of a garden (two acres in all) east of

[1] Conrad Gesner drew up a codified list of choice plants cultivated in the gardens of about twenty of his friends, with short lists of rarities in certain gardens. Johann Franke published his *Hortus Lusatiæ* in forty-eight pages—a very scarce work—which is a catalogue of all the plants growing near Launitz in Bohemia. The list contains both wild and cultivated plants, and the latter are distinguished by the addition of the letter H.

[2] This must have been Jean Robin, who in 1597 was appointed Keeper of the King's gardens in Paris. We know that Gerard was on intimate terms with him, and Robin sent him numerous plants, which he gratefully acknowledges in his Herbal. Gerard frequently speaks of him as "my loving friend John Robin."

Somerset House for four pence a year. Besides the rent he had to give " at the due and proper seasons of the yeare a convenient proportion and quantitie of herbes, floures, or fruite, renewing or growing within the said garden plott or piece of grounde, by the arte and industrie of the said John Gerard, if they be lawfully required and demanded." [1] Gerard only kept this garden for a year. In 1605 he parted with his interest in it to Robert Earl of Salisbury, and it is interesting to note that in the legal documents connected with this transaction he is described as herbarist to James I. Of his private life we know nothing beyond that he was married and that his wife helped him in his work. He died in February 1611–1612, and was buried in St. Andrew's Church, Holborn.

In 1597, as we have seen, Gerard published the Herbal which made him famous, but its history, as his critics point out, reflects little credit on the author. John Norton, the Queen's printer, had commissioned Dr. Priest, a member of the College of Physicians, to translate Dodoens's *Pemptades* from Latin into English. Priest died before he finished his work and the unfinished translation came somehow into Gerard's hands. Gerard altered the arrangement of the herbs from that of Dodoens to that of de l'Obel in his *Adversaria*, and of Priest's translation he merely says : " Dr. Priest, one of our London College, hath (as I heard) translated the last edition of Dodoens, which meant to publish the same, but being prevented by death his translation likewise perished." There are no fewer than 1800 illustrations in the Herbal, most of them taken from the same wood-blocks that Tabernæmontanus (Bergzabern) used for his *Eicones* (1590). Norton, the Queen's printer, procured the loan of these wood-blocks from Nicolas Bassæus of Frankfurt. They are good specimens, and certainly superior to the sixteen original cuts which Gerard added. It is interesting, however, to note that amongst the latter is the first published representation of the " Virginian " potato. Gerard made so many mistakes

[1] MSS. Record Office, James I. (Domestic), Vol. IX. fol. 113.

in connection with the illustrations that James Garret, a London apothecary (and the correspondent of Charles de l'Escluse), called Norton's attention to the matter. Norton thereupon asked de l'Obel to correct the work, and, according to de l'Obel's own account, he was obliged to make over a thousand alterations. Gerard then stopped any further emendation, on the ground that the work was sufficiently accurate, and declared further that de l'Obel had forgotten the English language. Mr. B. D. Jackson affirms that when one compares the Herbal with the catalogue of the plants in his garden Gerard seems to have been in the right. On the other hand, de l'Obel in his *Illustrationes* speaks of Gerard with great bitterness and alleges that the latter pilfered from the *Adversaria* without acknowledgment.

When one turns to the Herbal one forgets the bitterness of these old quarrels and Gerard's possible duplicity in the never-failing charm of the book itself. It is not merely a translation of Dodoens's *Pemptades*, for throughout the volume are inserted Gerard's own observations, numerous allusions to persons and places of antiquarian interest, and a good deal of contemporary folk-lore. No fewer than fifty of Gerard's own friends are mentioned, and one realises as one wanders through the pages of this vast book that he received plants from all the then accessible parts of the globe. Lord Zouch sent him rare seeds from Crete, Spain and Italy. Nicholas Lete, a London merchant, was a generous contributor to Gerard's garden and his name appears frequently. Gerard writes of him: "He is greatly in love with rare and faire flowers, for which he doth carefully send unto Syria, having a servant there at Aleppo, and in many other countries." It was Nicholas Lete who sent Gerard an "orange tawnie gilliflower" from Poland. William Marshall, a chirurgeon on board the *Hercules*, sent him rarities from the Mediterranean. The names which appear most frequently in connection with indigenous plants are those of Thomas Hesketh, a Lancashire gentleman, Stephen Bridwell, "a learned and diligent searcher of simples in the West of England," James

PORTRAIT OF JOHN GERARD FROM THE FIRST EDITION OF THE "HERBALL" (1597)

Cole, a London merchant, "a lover of plants and very skilful in the knowledge of them," and James Garret, a London apothecary and a tulip enthusiast, who "every season bringeth forth new plants of sundry colours not before seen, all of which to describe particularly were to roll Sisiphus's stone or number the sands." Jean Robin, the keeper of the royal gardens in Paris, sent him many rarities. For instance, of barrenwort (*Epimedium alpinum*) he writes: "This was sent to me from the French King's herbarist Robinus dwellying in Paris at the syne of the blacke heade in the street called Du bout du Monde, in English the end of the world." In view of Sir Walter Raleigh's well-known enthusiasm for collecting rare plants, it is at least possible that he may have been a donor to Gerard's garden.

Even the most cursory reading of the book suggests how much we lose by the lack of the old simple belief in the efficacy of herbs to cure not only physical ills, but also those of the mind and even of the heart. This belief was shared by the greatest civilisations of antiquity, and it is only we poor moderns who ignore the fact that "very wonderful effects may be wrought by the Vertues, which are enveloped within the compasse of the Green Mantles wherewith many Plants are adorned."[1] Doctors are cautious folk nowadays, but it is wonderful to think of a time when the world was so young that people were brave and hopeful enough to imagine that mere humans could alleviate, even cure, the sorrows of others. If ever anything so closely approaching the miraculous is attempted again one feels very sure that we shall turn, as the wise men of the oldest civilisations did, to God's most beautiful creations to accomplish the miracle. In common with the majority of the old herbalists, Gerard had a faith in herbs which was simple and unquestioning. Sweet marjoram, he tells us, is for those "who are given to over-much sighing." Again, "The smell of Basil is good for the heart . . . it taketh away sorrowfulness, which commeth of melancholy and maketh a man merry and glad." "Bawme

[1] W. Coles, *The Art of Simpling*.

comforts the heart and driveth away all melancholy and sadnesse : it makes the heart merry and joyfull and strengtheneth the vitall spirits." "Chervil root boiled and after dressed as the cunning Cook knoweth how better than myself is very good for old people that are dull and without courage." Of the despised dead-nettle he tells us that "the flowers baked with sugar, as roses are, maketh the vitall spirits more fresh and lively." In connection with borage he quotes the well-known old couplet :

"I Borage
Bring alwaies Courage."

"Those of our time," he continues, "do use the floures in sallads to exhilerate the mind and make the mind glad. There be also many things made of them, used everywhere for the comfort of the heart, for the driving away of sorrow and encreasing the joy of the minde. . . . The leaves and floures put into wine make men and women glad and merry and drive away all sadnesse, dulnesse and melancholy."

Of bugloss he says : "The physitions use the leaves, floures and rootes and put them into all kindes of medecines indifferently, which are of force and vertue to drive away sorrow and pensiveness of the minde, and to comfort and strengthen the heart."

Rosemary was held of such sovereign virtue in this respect that even the wearing of it was believed to be remedial. "If a garland thereof be put about the head, it comforteth the brain, the memorie, the inward senses and comforteth the heart and maketh it merry." Certain herbs strewed about the room were supposed to promote happiness and content. Meadowsweet, water-mint and vervain (one of the three herbs held most sacred by the Druids) were those most frequently used for this purpose.

"The savor or smell of the water-mint rejoyceth the heart of man, for which cause they use to strew it in chambers and places of recreation, pleasure and repose, where feasts and banquets are made."

"The leaves and floures of meadowsweet farre excelle all

other strowing herbs for to decke up houses, to strawe in chambers, halls and banqueting houses in the summertime, for the smell thereof makes the heart merrie and joyful and delighteth the senses."

In connection with vervain he quotes Pliny's saying that "if the dining room be sprinckled with water in which the herbe hath been steeped the guests will be the merrier."

Scattered through the Herbal we find recipes for the cure of many other ailments with which modern science does not attempt to cope. For instance, under "peony" we read: "The black graines (that is the seed) to the number of fifteene taken in wine or mead is a speciall remedie for those that are troubled in the night with the disease called the Night Mare, which is as though a heavy burthen were laid upon them and they oppressed therewith, as if they were overcome with their enemies, or overprest with some great weight or burthen, and they are also good against melancholie dreames." Under Solomon's seal one lights on this: "The root stamped while it is fresh and greene and applied taketh away in one night or two at the most any bruise, black or blew spots, gotten by falls or women's wilfulnesse in stumbling upon their hasty husbands' fists or such like." Of cow parsnip he tells us: "If a phrenticke or melancholicke man's head bee anointed with oile wherein the leaves and roots have been sodden, it helpeth him very much, and such as bee troubled with the sickness called the forgetfull evill." Would any modern have either the courage or the imagination to attempt to cure "the forgetfull evill"? In the old Saxon herbals the belief in the efficacy of herbs used as amulets is a marked feature, and even in Gerard's Herbal much of this old belief survives. "A garland of pennyroyal," he tells us, "made and worne about the head is of a great force against the swimming in the head, the paines and giddiness thereof." The root of spatling poppy "being pound with the leaves and floures cureth the stinging of scorpions and such like venemous beasts: insomuch that whoso doth hold the same in his hand can receive no damage or hurt by any venemous beast." Of shrubby trefoil we learn that "if a man hold it in his hand he cannot be hurt with the biting of any venemous beast." Of rue he says: "If a man be anointed with the juice of rue, the poison of wolf's bane, mushrooms or todestooles, the biting of serpents, stinging of scorpions, spiders, bees, hornets and wasps will not hurt him." In the older herbals numerous herbs are mentioned as being of special virtue when used as amulets to protect the wayfaring man from weariness, but Gerard mentions only two—mugwort and *Agnus castus*. He quotes the authority of Pliny for the belief that "the traveller or wayfaring man that hath mugwort tied about him feeleth no wearisomeness at all and he who hath it about him can be hurt by no poysonous medecines, nor by any wilde beaste, neither yet by the Sun itselfe." Of *Agnus castus* he writes: "It is reported that if such as journey or travell do carry with them a branch or rod of agnus castus in their hand, it will keep them from weariness." The herbs most in repute as amulets against misfortune generally were angelica (of sovereign virtue against witchcraft and enchantments) and figwort, which was "hanged about the necke" to keep the wearer in health. At times one feels that Gerard rather doubted the efficacy of these "physick charms," and he gives us a naïve description of his friends' efforts to cure him of an ague by their means.

"Having a most grievous ague," he writes, "and of long continuance, notwithstanding Physick charmes, the little wormes found in the heads of Teazle hanged about my necke, spiders put in a walnut shell, and divers such foolish toies, that I was constrained to take by fantasticke peoples procurement, notwithstanding I say my helpe came from God himselfe, for these medicines and all other such things did me no good at all."

Under "gourd" Gerard gives a use of this herb which, though popular, is not to be found in any other English herbal. "A long gourd," he says, "or else a cucumber being laid in the cradle or bed by the young infant while it is asleep and sicke of an ague, it shall very quickly be made whole." The cure was presumably effected by the cooling properties of the fruit. In another place he recommends the use of branches of willow for a similar purpose. "The greene boughes of willows with the leaves may very well be brought into chambers and set about the beds of those that be sick of fevers, for they do mightily coole the heate of the aire, which thing is wonderfull refreshing to the sicke Patient."

There is so much contemporary folk lore embodied in Gerard that it is disappointing to find that when writing of mugwort, a herb which has been endowed from time immemorial with wonderful powers, he declines to give the old superstitions "tending to witchcraft and sorcerie and the great dishonour of God; wherefore do I purpose to omit them as things unwoorthie of my recording or your receiving." He also pours scorn on the mandrake legend. "There have been," he says, "many ridiculous tales brought up of this plant, whether of old wives or runnegate surgeons, or phisick mongers I know not, all whiche dreames and old wives tales you shall from hencefoorth cast out of your bookes of memorie." The old legend of the barnacle geese, however, he gives fully. It is both too long and too well known to quote, but it is interesting to remember that this myth is at least as old as the twelfth century. According to one version, certain trees growing near the sea produced fruit like apples, each containing the embryo of a goose, which, when the fruit was ripe, fell into the water and flew away. It is, however, more commonly met with in the form that the geese emanated from a fungus growing on rotting timber floating at sea. This is Gerard's version. One of the earliest mentions of this myth is to be found in Giraldus Cambrensis (*Topographia Hiberniæ*, 1187), a zealous reformer of Church abuses. In his protest against eating these barnacle geese during Lent he writes thus :—

"There are here many birds which are called Bernacae

which nature produces in a manner contrary to nature and very wonderful. They are like marsh geese but smaller. They are produced from fir-timber tossed about at sea and are at first like geese upon it. Afterwards they hang down by their beaks as if from a sea-weed attached to the wood and are enclosed in shells that they may grow the more freely. Having thus in course of time been clothed with a strong covering of feathers they either fall into the water or seek their liberty in the air by flight. The embryo geese derive their growth and nutriment from the moisture of the wood or of the sea, in a secret and most marvellous manner. I have seen with my own eyes more than a thousand minute bodies of these birds hanging from one piece of timber on the shore enclosed in shells and already formed . . . in no corner of the world have they been known to build a nest. Hence the bishops and clergy in some parts of Ireland are in the habit of partaking of these birds on fast days without scruple. But in doing so they are led into sin. For if anyone were to eat the leg of our first parent, although he (Adam) was not born of flesh, that person could not be adjudged innocent of eating flesh."

Jews in the Middle Ages were divided as to whether these barnacle geese should be killed as flesh or as fish. Pope Innocent III. took the view that they were flesh, for at the Lateran Council in 1215 he prohibited the eating of them during Lent. In 1277 Rabbi Izaak of Corbeil forbade them altogether to Jews, on the ground that they were neither fish nor flesh. Both Albertus Magnus and Roger Bacon derided the myth, but in general it seems to have been accepted with unquestioning faith. Sebastian Munster, in his *Cosmographia Universalis* (1572), tells us that Pope Pius II. when on a visit to Scotland was most anxious to see these geese, but was told that they were to be found only in the Orkney Islands. Sebastian believed in them himself, for he wrote of them :—

"In Scotland there are trees which produce fruit conglomerated of leaves, and this fruit when in due time it falls into the water beneath it is endowed with new life and is converted into a living bird which they call the tree-goose. . . . Several old cosmographers, especially Saxo Grammaticus, mention the tree and it must not be regarded as fictitious as some new writers suppose."[1]

Even Hector Boece, in his *Hystory and Croniklis of Scotland* (1536), took the myth seriously, but in his opinion "the nature of the seis is mair relevant caus of their procreation than ony uther thyng." William Turner accepted the myth and gives as his evidence what had been told him by an eye-witness, "a theologian by profession and an Irishman by birth, Octavian by name," who promised him that he would take care that some growing chicks should be sent to him! In later times we find that Gaspar Schott (*Physica Curiosa Sive Mirabilia Naturae et Artis*, 1662, lib. ix. cap. xxii. p. 960) quotes a vast number of authorities on the subject and then demonstrates the absurdity of the myth. Yet in 1677 Sir Robert Moray read before the Royal Society "A Relation concerning Barnacles," and this was published in the *Philosophical Transactions*, January–February 1677–8. Among illustrations of the barnacle geese, that in de l'Obel's *Stirpium Historia* (1571) depicts the tree without the birds. Gerard shows the tree with the birds; in Aldrovandus leaves have been added to the tree and there is also an illustration showing the development of the barnacles into geese.

As in all herbals the element of the unexpected is not lacking in Gerard. Who would think of finding under the eminently dull heading "fir trees" the following gem of folk lore? "I have seen these trees growing in Cheshire and Staffordshire and Lancashire, where they grew in great plenty as is reported before Noah's flood, but then being overturned and overwhelmed have lien since in the mosse and waterie moorish grounds very fresh and sound untill this day; and so full of a resinous substance, that they burne like a Torch or Linke and the inhabitants of those countries do call it Fir-wood and Fire-wood unto this

[1] *Cosmographia Universalis*, 1572, p. 49.

day: out of the tree issueth the rosin called Thus, in English Frankincense." In these days of exaggerated phraseology one is the more appreciative of that word "overturned." Gerard mentions the famous white Thorn at Glastonbury, but he is very cautious in his account of it. "The white thorn at Glastonbury . . . which bringeth forth his floures about Christmas by the report of divers of good credit, who have seen the same; but myselfe have not seen it and therefore leave it to be better examined."

Another attractive feature of this Herbal is the preservation in its pages of many old English names of plants. One species of cudweed was called "Live-for-ever." "When the flower hath long flourished and is waxen old, then comes there in the middest of the floure a certain brown yellow thrumme, such as is in the middest of the daisie, which floure being gathered when it is young may be kept in such manner (I meane in such freshnesse and well-liking) by the space of a whole year after in your chest or elsewhere; wherefore our English women have called it 'Live-long,' or 'Live-for-ever,' which name doth aptly answer his effects." Another variety of cudweed was called "Herbe impious" or "wicked cudweed," a variety "like unto the small cudweed, but much larger and for the most part those floures which appeare first are the lowest and basest and they are overtopt by other floures, which come on younger branches, and grow higher as children seeking to overgrow or overtop their parents (as many wicked children do), for which cause it hath been called 'Herbe impious.'" Of the herb commonly known as bird's-eye he tells us: "In the middle of every small floure appeareth a little yellow spot, resembling the eye of a bird, which hath moved the people of the north parts (where it aboundeth) to call it Birds eyne." "The fruitful or much-bearing marigold," he writes, "is likewise called Jackanapes-on-horsebacke: it hath leaves, stalkes and roots like the common sort of marigold, differing in the shape of his floures; for this plant doth bring forth at the top of the stalke one floure like the other marigolds, from which start forth sundry other smal floures, yellow likewise and of the same fashion as the first, which if I be not deceived commeth to pass per accidens, or by chance, as Nature often times liketh to play with other floures; or as children are borne with two thumbes on one hand or such like, which living to be men do get children like unto others: even so is the seed of this marigold, which if it be sowen it brings forth not one floure in a thousand like the plant from whence it was taken." Goat's-beard still retains its old name of 'go-to-bed-at-noon,' "for it shutteth itselfe at twelve of the clocke, and sheweth not his face open untill the next dayes Sun doth make it flower anew, whereupon it was called go-to-bed-at-noone: when these floures be come to their full maturitie and ripenesse they grow into a downy Blow-ball like those of dandelion, which is carried away with the winde." Of the wild scabious (still called devil's-bit by country folk) he tells us: "It is called Devil's bit of the root (as it seemeth) that is bitten off. Old fantasticke charmers report that the Devil did bite it for envie because it is an herbe that hath so many good vertues and is so beneficent to mankind." Gerard's, again, is the only herbal in which we find one of the old names for vervain: "Of some it is called pigeons grasse because Pigeons are delighted to be amongst it as also to eat thereof." Golden moth-wort, he tells us, is called God's flower "because the images and carved gods were wont to wear garlands thereof: for which purpose Ptolomy King of Egypt did most diligently observe them as Pliny writeth. The floures . . . glittering like gold, in forme resembling the scaly floures of tansy or the middle button of the floures of camomil, which, being gathered before they be ripe or withered, remaine beautiful long after, as myself did see in the hands of Mr. Wade, one of the Clerks of her Majesties Counsell, which were sent him among other things from Padua in Italy." The variety of daisy which children now call "Hen and Chickens" was known as the "childing daisy" in Gerard's time. "Furthermore, there is another pretty double daisy which differs from the first described only in the floure which at the sides thereof puts forth many foot-stalkes carrying also little double floures, being commonly of a red colour; so that each stalke carries as it were an old one and the brood thereof: whence they have fitly termed it the childing Daisie." Of silverweed he tells us: "the later herbarists doe call it argentine of the silver drops that are to be seen in the distilled water thereof, when it is put into a glasse, which you shall easily see rowling and tumbling up and downe in the bottome." Delphinium, we learn, derives its name from dolphin, "for the floures especially before they be perfected have a certain shew and likeness of those Dolphines which old pictures and armes of certain antient families have expressed with a crooked and bending figure or shape, by which signe also the heavenly Dolphin is set forth." Rest-harrow, he says, is so called "because it maketh the Oxen whilest they be in plowing to rest or stand still." One of the most attractive names which he accounts for is cloudberry. "Cloudberrie groweth naturally upon the tops of two high mountaines (among the mossie places), one in Yorkshire, called Ingleborough, the other in Lancashire called Pendle, two of the highest mountains in all England, where the clouds are lower than the tops of the same all winter long, whereupon the country people have called them cloud-berries; found there by a curious gentleman in the knowledge of plants, called Mr. Hesketh, often remembered."

For those who care to seek it Gerard supplies an unequalled picture of the wild-flower life in London in Elizabethan days. It is pleasant to think of the little wild bugloss growing "in the drie ditch bankes about Piccadilla" (Piccadilly), of mullein "in the highwaies about Highgate"; of clary "in the fields of Holborne neere unto Grays Inn"; of lilies of the valley, the rare white-flowered betony, devil's-bit, saw-wort, whortle-berries, dwarf willows and numerous other wild plants on Hampstead Heath; of the yellow-flowered figwort "in the moist medowes as you go from London to Hornsey"; of the yellow pimpernel "growing in abundance between Highgate and Hampstead"; of sagittaria "in the Tower ditch at London"; of white saxifrage "in the great field by Islington called the Mantles and in Saint George's fields behinde Southwarke"; of the vervain mallow "on the ditch sides on the left hand of the place of execution by London called Tyburn and in the bushes as you go to Hackney"; of marsh-mallows "very plentifully in the marshes by Tilbury Docks"; of the great wild burnet "upon the side of a causey, which crosseth a field whereof the one part is earable ground and the other part medow, lying between Paddington and Lysson Green neere unto London upon the highway"; of hemlock dropwort "betweene the plowed lands in the moist and wet furrowes of a field belonging to Battersey by London, and amongst the osiers against York House a little above the Horse-ferry against Lambeth"; of the small earth-nut "in a field adjoyning to Highgate on the right side of the middle of the village and likewise in the next field and by the way that leadeth to Paddington by London"; of chickweed spurry "in the sandy grounds in Tothill fields nigh Westminster"; of the pimpernel rose "in a pasture as you goe from a village hard by London called Knightsbridge unto Fulham, a village thereby"; of dwarf elder "in untoiled places plentifully in the lane at Kilburne Abbey by London"; of silver cinquefoil "upon brick and stone walls about London, especially upon the bricke wall in Liver Lane"; of water-ivy, "which is very rare to find, nevertheless I found it once in a ditch by Bermondsey house near to London and never elsewhere."

The glimpses he gives us of London gardens are few and one longs for more. It is remarkable how few vegetables, or "potherbs" as they called them, were grown in Elizabethan times. Vegetables which figured in the old Roman menus were considered luxuries in this country even in the days of the later Stuarts. The wild carrot is an indigenous plant in our islands, but of the cultivated carrot we were ignorant till the Flemish immigrants introduced it in the early seventeenth century. Parsnips, turnips and spinach were also rarities. With the exception of the wild cabbage, the whole brassica tribe were unknown to us till the late sixteenth and seventeenth centuries. Potatoes and Jerusalem artichokes were both introduced into this country in Tudor days. Gerard was one of the first to grow potatoes, and he proudly tells us, "I have received hereof from Virginia roots which grow and prosper in my garden as in their own native countrie." He was, in fact, the originator of the popular but incorrect epithet "Virginia potato." The potato was not a native of Virginia, nor was it cultivated there in Tudor times. The Spaniards brought it from Quito in 1580, and Gerard had it in his garden as early as 1596. The potato to which Shakespeare refers (*Troilus and Cressida*, V. ii. 534; *Merry Wives of Windsor*, V. v. 20, 21) is, of course, the sweet potato, which had been introduced into Europe nearly eighty years earlier. Gerard speaks of this sweet potato as "the common potato," which is somewhat confusing to the modern reader.

There is a delightful glimpse of a well-known London garden, that of "Master Tuggie," who lived in Westminster and whose hobby was gilliflowers. It is the more interesting to find this passage in Gerard, for, as all lovers of Parkinson's *Paradisus* will remember, some of the varieties of gilliflower were called after their enthusiastic grower. Indeed, who can forget their enchanting names—"Master Tuggie's Princesse" and "Master Tuggie his Rose gillowflower"? Of gilliflowers, which vied with roses in pride of place in Elizabethan gardens, Gerard writes thus :—

"Now I (holding it a thing not so fit for me to insist upon these accidental differences of plants having specifique differences enough to treat of) refer such as are addicted to these commendable and harmless delights to survey the late and oft-mentioned Worke of my friend, Mr. John Parkinson, who hath accurately and plentifully treated of these varieties. If they require further satisfaction, let them at the time of the yeare repaire to the garden of Mistress Tuggie (the wife of my late deceased friend, Mr. Ralph Tuggie) in Westminster, which in the excellencie and varietie of these delights exceedeth all that I have seene, as also, he himself, whilst he lived exceeded most, if not all, of his time, in his care, industry and skill, in raising, increasing and preserving of these plants."

Gerard's descriptions of the most loved English garden flowers are perhaps too well known to quote, and therefore I give only the following: "The Plant of Roses, though it be a shrub full of prickes, yet it hath beene more fit and convenient to have placed it with the most glorious flowers of the world than to inserte the same here among base and thornie shrubs; for the rose doth deserve the chiefest and most principall place among all flowers whatsoever being not only esteemed in his beautie, vertue and his fragrance and odoriferous smell, but also because it is the honor and ornament of our English Scepter, as by the coniunction appeereth in the uniting of those two most royal houses of Lancaster and Yorke. Which pleasant flowers deserve the chiefest place in crowns and garlands. The double white sort doth growe wilde in many hedges of Lancashire in great abundance, even as briers do with us in these southerly parts, especially in a place of the countrey called Leyland, and in the place called Roughfoorde not far from Latham. The distilled water of roses is good for the strengthening of the hart and refreshing of the spirits and likewise in all things that require a gentle cooling. The same being put in iunketting dishes, cakes, sawces and many other pleasant things, giveth a fine and delectable taste. It bringeth sleepe which also the fresh roses themselves promote through their sweete and pleasant smell."

Like most gardeners Gerard was an optimist. It is wonderful enough to think of the rare, white thyme growing in the heart of London, but think of the courage of trying to raise dates in the open ! "of the which," Gerard tells us (in no wise downcast by his numerous failures), "I have planted many times in my garden and have growne to the height of three foot, but the frost hath nipped them in such sort that soone after they perished, notwithstanding my industrie by covering them, or what else I could do for their succour." And does it not make one feel as eager as Gerard himself when one finds, under water-mallows, that, though exotic plants, "at the impression hereof I have sowen some seeds of them in my garden, expecting the successe." The mere catalogue of the plants in Gerard's own wonderful garden fills a small book, and scattered through the Herbal we find numerous references to it, unfortunately too lengthy to quote here.

One likes to think that Shakespeare must have seen this garden, for we know that at least for a time he lived in the vicinity. In those days two such prominent men could scarcely have failed to know one another.[1] As Canon Ellacombe has pointed out, Shakespeare's writings are full of the old English herb lore. In this use of plant lore, which was traditional rather than literary, he is curiously distinct from his contemporaries. Outside the herbals there is more old English herb lore to be found in Shakespeare than in any other writer. It is, in fact, incredible that the man whose own works are so redolent of the fields and hedgerows of his native Warwickshire, did not visit the garden of the most famous herbarist of his day. Perhaps it was to Shakespeare that Gerard first told the sad tale of the loss of his precious scammony of Syria, a tale which no one with a gardener's heart can read without a pang of sympathy, even

[1] Shakespeare and Gerard were near neighbours during the time when the former was writing many of his finest plays, for Shakespeare lived in the house of a Huguenot refugee (Mountjoy by name) 1598–1604. This house was at the corner of Mugwell Street (now Monkswell Street) and Silver Street, very near the site of the ancient palace of King Athelstan in Saxon days. Almost opposite Mountjoy's house was the Barber-Surgeons' Hall. Aggers' Map (*circ.* 1560) with pictures of the houses, gives an excellent idea of the neighbourhood in those days. See also Leak's Map (1666).

after the lapse of three centuries. One of his numerous correspondents had sent him the seed of this rare plant, "of which seed," he says:—

"I received two plants that prospered exceeding well; the one whereof I bestowed upon a learned apothecary of Colchester, which continueth to this day bearing both floures and ripe seed. But an ignorant weeder of my garden plucked mine up and cast it away in my absence instead of a weed, by which mischance I am not able to write hereof so absolutely as I determined. It floured in my garden about S. James' tide as I remember, for when I went to Bristow Faire I left it in floure; but at my returne it was destroyed as is aforesaid."

CHAPTER V

HERBALS OF THE NEW WORLD

" And I doe wish all Gentlemen and Gentlewomen, whom it may concerne, to bee as careful whom they trust with the planting and replanting of these fine flowers, as they would be with so many Iewels."—PARKINSON, *Paradisus*, 1629.

To English folk and Americans alike the herbals—now amongst the rarest in the English language—treating of the virtues of herbs in the New World are of exceptional interest. For these contain some of the earliest records of the uses of herbs learnt from the Red Indians, lists of English weeds introduced into America by the first settlers, and, perhaps most interesting of all, what they grew in the first gardens in New England. It requires very little imagination to realise how much the discovery of the New World meant to the botanists, gardeners and herbalists of that day, for at no time in our history were there greater plant-lovers than in Elizabethan and Stuart times. In their strenuous lives the soldiers, explorers and sea-captains found time to send their friends in the Old World rare plants and other treasures, and these gifts of " rarities " were cherished as jewels. Is not the following a vivid picture of the arrival of such a package from the New World? " There came a Paket, as of Letters, inrolled in a seare clothe : so well made that thei might passe to any part beeyng never so farre, the whiche beeyng opened, I founde a small Cheste made of a little peece of Corke, of a good thickenesse sette together, whiche was worthie to be seen, and in the holownesse of it came the hearbes, and the seedes that the Letter speaketh of, everythyng written what it was, and in one side of the Corke, in a hollowe place there came three Bezaar stones, cloased with a Parchement and with Waxe in good order. The Letter was written with verie small Letters, and sumwhat harde to reade." The letter and the precious gift of herbs, seeds and stones were from an officer on duty in "New Spain" (he describes himself as "a Souldier that have followed the warres in these countries all my life"), who was unknown to Monardes, but had read his first book on the use of the herbs in the New World, and therefore was emboldened to send him these rare plants and the "bezaar stones." Nicolas Monardes, the author of this herbal (translated into English by John Frampton),[1] most gratefully acknowledges his unknown friend's kindness and writes of him, "the gentleman of the Peru, which wrote to me this letter, although I know hym not, it seemeth that he is a man curious and affectioned to the like thinges and I have him in great estimation. For bicause that the office of a Souldier is to handle weapons, and to sheed bloud, and to do other exercises apertainyng to Souldiers, he is muche to bee esteemed that he will enquire and searche out herbes, and Plantes and to knowe their properties and vertue. And therefore I dooe esteeme muche of this Gentlemanne for the labour whiche he taketh in knowyng and enquiryng of these naturall thinges. And I doe owe much unto him, ... I wil provoke hym by writyng to hym againe, to sende more thinges. For it is a greate thinge to knowe the secreates and marvailes of nature, of the Hearbes which he hath sent me. I will make experience of them and I will know their vertues and operation and the Seedes wee will sowe at their time."

The interest in the plant-life of the New World may be judged from the fact that Monardes's work, which is the earliest "American" herbal, was translated into Latin by no less a botanist than Charles de l'Escluse, and into Italian, Flemish, French and English. Frampton's English translation went through four editions. The original book was written nineteen

[1] *Joyfull Newes out of the newe founde worlde wherein is declared the rare and singular vertues of diuerse and sundrie Hearbes, etc.* See Bibliography of English Herbals, p. 211. Nicolas Monardes was a Spanish doctor living in Seville and his book was written in 1569 (see p. 231).

years before the defeat of the Spanish Armada, and throughout it there is very evident the pride of a Spanish subject in the splendid overseas dominions of his country, then the first empire in the world and the mistress of the seas. The preface is so redolent of the atmosphere of Spanish galleons and the boundless interest in the great new continent and its wonders, that I quote it almost in full, although in modern print it loses much of the charm of the original black-letter. The writer surely had in his mind the account of the navy of Tharshish, which came once in three years, " bringing gold and silver, ivory, and apes, and peacocks," and one cannot help suspecting that loyalty to his Catholic Majesty of Spain suggested the inclusion of lions from America in order that he might not be outdone by the splendour of Solomon. Moreover, Monardes proudly tells us that from the New World to Spain " there commeth every yere one hundred shippes laden . . . that it is a greate thynge and an incredle riches."

" In the yere of our Lorde God, a thousande, fower hundreth ninetie twoo : our Spaniardes were gouerned by Sir Christofer Colon [Columbus], beeyng naturally borne of the countrie Genoa, for to discouer the Occidentall Indias, that is called at this daie, the Newe Worlde, and thei did discouer the first lande thereof, the XI daie of October, of the saied yere, and from that tyme unto this, thei haue discouered many and sundrie Ilandes, and muche firme Lande, as well in that countrie, whiche thei call the Newe Spaine, as in that whiche is called the Peru, where there are many Prouinces, many Kyngdomes, and many Cities, that hath contrary and diuers customes in them, whiche there hath been founde out, thynges that neuer in these partes, nor in any other partes of the worlde hath been seen, nor unto this daie knowen : and other thynges, whiche now are brought unto us in greate aboundaunce, that is to saie, Golde, Siluer, Pearles, Emeraldes, Turkeses [turquoises], and other fine stones of greate value, yet greate is the excesse and quantitie that hath come, and every daie doeth come, and in especiallie of Golde and Siluer: That it is a thyng of admiration that the greate number of Milleons, whiche hath come besides the greate quantitie of Pearles, hath filled the whole worlde, also thei doe bryng from that partes, Popingaies, Greffons, Apes, Lions, Gerfaucons, and other kinde of Haukes, Tigers wolle, Cotton wolle, Graine to die colours with all, Hides, Sugars, Copper, Brasill, the woode Ebano, Anill: and of all these, there is so greate quantitie, that there commeth every yere, one hundred Shippes laden thereof, that it is a greate thynge and an incredle riches.

"And besides these greate riches, our Occidentall Indias doeth sende unto us many Trees, Plantes, Herbes, Rootes, Joices, Gummes, Fruites, Licours, Stones that are of greate medicinall vertues, in the whiche there bee founde, and hath been founde in them, verie greate effectes that doeth excede muche in value and price: All that aforesaied, by so muche as the Corporall healthe is more Excellent, and necessaire then the temporall goodes, the whiche thynges all the worlde doeth lacke, the wante whereof is not a little hurtfull, according to the greate profite which wee doe see, by the use of them doeth followe, not onely in our Spaine but in all the worlde. . . . The people of old tyme did lacke them, but the tyme whiche is the discouerer of all thynges, hath shewed them unto us greatly to our profite, seying the greate neede that we had of them.

"And as there is discouered newe regions, newe Kyngdomes, and newe Prouinces, by our Spanyardes, thei haue brought unto us newe Medicines, and newe Remedies, wherewith thei doe cure and make whole many infirmities, whiche if wee did lacke them, thei were incurable, and without any remedie, the whiche thynges although that some have knowledge of them, yet thei bee not common to all people, for whiche cause I did pretēde to treate and to write, of all thynges, that thei bryng from our Indias, whiche serueth for the arte and use of Medicine, and the remedy of the hurtes and deseases, that wee doe suffer and endure, whereof no small profite doeth followe to those of our tyme, and also unto them that shall come after us, the whiche I shall be the first, that the rather the followers maie adde hereunto, with this beginnyng, that whiche thei shall more knowe, and by experience shall finde. And, as in this Citee of Seuill, which is the Porte and skale of all Occidentall Indias, wee doe knowe of thē more, then in any other partes of all Spaine, for because that all thynges come first hither, where with better relation, and greater experience it is knowen. I doe it with experience and use of them this fourtie yeres, that I doe cure in this Citee, where I haue informed myself of them, that hath brought these thynges out of those partes with muche care, and I have made with all diligence and foresight possible, and with much happie successe."

Then he begins straightway to tell us of various herbs and gums brought from the New World, and of what the herbalists had been able to learn of their medicinal virtues. He writes of "Copall" and "Anime" (varieties of rosin), and tells us that the Spaniards first learnt of these from the Indian priests, who "went out to receive them [the Spaniards] with little firepottes, burnyng in them this Copall, and giuing to them the smoke of it at their noses." "Tacamahaca" (the Indian name for a rosin) is "taken out by incision of a tree beyng as greate as a Willowe Tree, and is of a verie sweete smell; he doeth bryng forth a redde fruite, as the seede of Pionia." The Indians used it for swellings in any part of the body and also for toothache. "Caranna," another gum brought from Nombre de Dios, is discovered to be of sovereign virtue for gout—"it taketh it awaie with muche easines." The balsam of the New World, "that licour most excellent whiche for his Excellencie and meruerlous effectes is called Balsamo, an imitation of the true Balsamo that was in the lande of Egipt," is "made of a tree greater than a Powndgarned Tree, it carrieth leaues like to Nettles: the Indians doe call it Xilo and we do call the same

Balsamo." There follows an account of the way in which the Red Indians made the balsam, either by cutting incisions in the tree and letting the "clammish licour, of colour white but most excellent and very perfite," run out, or by cutting up boughs and branches of the tree into very small pieces, boiling them in cauldrons and then skimming off the oil. "It is not convenient, nor it ought to be kept in any other vessel then in silver (glasse or Tinne or any other thing glassed, it doth penetrate and doth passe through it), the use thereof is onely in thinges of Medecine and it hath been used of long tyme . . . the Spaniards had knowledge of it because they did heale therwith the woundes that they did receive of the Indians: beyng advised of the vertue thereof by the same Indians, and they did see the saide Indians heale and cure themselves therewith." We learn that when this precious new balsam was first brought to Spain it sold for ten ducats an ounce, and in Italy for a hundred ducats an ounce. The use of another wound herb, "for shottes of arrows," of which unfortunately he does not give even the Indian name, was taught to a certain "Jhon Infante" by his native servant. The book gives us many pleasant glimpses of the kindly courtesy of the Red Indians to their foes, and though, according to some authorities, they would never tell the secrets of the herbs they used as medicines, we have Monardes's detailed accounts of how they showed the Spaniards the uses of them. Guiacum, for instance, was brought to the notice of a Spaniard in San Domingo by an Indian doctor.

One of the most interesting accounts is of "Mechoacan." "It is brought from a countrie that is beyonde the greate Citie of Mexico more than fortie leagues, that is called Mechoacan, the whiche Syr Fernando Curtes did conquer in the yere of 1524, it is a countrie of muche Riches, of Gold and chiefly of Silver . . . those Mynes be so celebrated and of so muche riches that they be called the Cacatecas, every day they goe discovering in the Lande verie riche Mynes of Silver and some of Golde, it is a countrie of good and holsome ayres, and doth bring forth healthfull Hearbes for to heale many diseases, in so muche that at the tyme the Indians had the government of it, the inhabiters there rounde aboute that Province, came thether to heale their diseases and infirmities. . . . The Indians of that countrie be of a taller growthe and of better faces then the Borderers are and of more healthe.

"The principall place of that province the Indians doe call in their language Chincicila and the Spaniards doe call it as thei call that realme Mechoacan, and it is a great towne of Indians, situated nere to a lake which is of swete water and of verie muche Fishe, the same Lake is like the fashion of making an horse shewe, and in the middest thereof standeth the Towne, the whiche at this daye hath greate trade of buying and sellyng."

We are told in detail how the Warden of the Friars of St. Francis was cured by a native Indian doctor with this herb—"mechoacan":—

"As soone as that Province was gotten of the Indians there went thither certaine Friers, of Saincte Frances order, and as in a countrie so distant from their naturall soyle, some of them fell sicke, amongest whom the Warden, who was the Chief Frier of the house fell sicke, with whom Caconcin Casique, an Indian lord, a man of great power in that countrie, had very greate friendship, who was Lorde of all that countrie. The father Warden had a long sicknes and put to muche danger of life, the Casique as he sawe his disease procede forward, he saied that he would bryng hym an Indian of his, which was a Phisition, with whom he did cure hym self, and it might bee that he would give hym remeady of his disease. The whiche beeyng heard of the Frier, and seyng the little helpe that he had there, and the want of a Phisition, and other thynges of benefite, he thanked hym and saied unto hym, that he should bryng hym unto hym: who beyng come, and seyng his disease, he said to the Casique, that if he tooke a pouder that he would giue hym of a roote, that it would heale hym. The whiche beeyng knowen to the Frier, with the desire that he had of healthe, he did accepte his offer and tooke the pouder that the Indian Physition gave hym the nexte daie in a little Wine. . . . He was healed of his infirmitie and the rest of the Friers which were sicke did followe the father Warden's cure and took of the Self same powder once or twice and as ofte as thei had neede of for to heale them. The use of the whiche went so well with them that the Friers did send relation of this to the Father Provincall to Mexico where he was: who did communicate with those of the countrie, giving to them of the roote, and comforting them that thei should take it, because of the good relation that he had from those Friers of Mechoacan. The whiche beyng used of many and seyng the marueilous woorkes that it did the fame of it was extended all abrode, that in short tyme all the countrie was full of his good woorkes and effectes, banishing the use of Ruibarbe of Barbarie and taking his name, naming it Ruibarbo of the Indos and so all men dooeth commonly call it. And also it is called Mechoaca for that it is brought from thence. . . . And so thei do carry it from the Newe Spaine as Merchandise of very great price."

The plant itself Monardes describes thus:—" It is an herbe that goeth creepyng up by certaine little Canes, it hath a sadde greene coulour, he carrieth certaine leaues, that the greatnesse of them maie bee of the greatnesse of a good potenge dishe, that is in compasse rounde, with a little point, the leaffe hath his little Senewes, he is small, well nere without moisture, the stalke is of the coulour of a cleare Taunie. Thei saie that he dooeth caste certaine clusters, with little Grapes, of the greatnesse of a Coriander seede, whiche is his fruite and dooeth waxe ripe by the Monethe of September: he doeth caste out many bowes, the whiche doeth stretche a long upō the yearth, and if you doe put anythyng nere to it, it goeth creepyng upon it. The roote of the Mechoacan is unsaverie and without bightyng or any sharpness of taste."

The book was published in successive parts, and the second

of these, dedicated to the King of Spain, contains the first written account and illustration of " the hearbe tabaco." Monardes tells us that this herb was one " of much antiquity " amongst the Indians, who taught the Spaniards to use it as a wound-herb. It was first introduced into Spain " to adornate Gardens with the fairenesse thereof and to give a pleasant sight, but nowe we doe use it more for his meruelous medicinable vertues than for his fairenesse." The Red Indians called it " picielt." (The name tabaco was given it by the Spaniards, either from the island which still bears the name Tobago, as Monardes declares, or from a native word connected in some way with the use of the dried leaves for smoking.) According to Monardes the leaves, when warmed and laid on the forehead with orange oil, were efficacious to cure headaches. They were also good for toothache. " When the griefe commeth of a cold cause or of colde Rumes, putting to the tooth a little ball made of the leafe of the Tabaco, washing first the tooth with a smal cloth wet in the Juyce, it stayeth it, that the putrifaction goe not forwarde : and this remedie is so common that it healeth euerie one." Of greater interest is the account of its application as a wound-herb and of an experiment made on a small dog at the Spanish Court.

" A little whiles past, certain wilde people going in their Bootes [boats] to S. John De puerto Rico to shoote at Indians or Spaniards (if that they might find them) came to a place and killed certain Indians and Spaniards and did hurt many, and as by chance there was no Sublimatum at that place to heale them, they remembered to lay upon the wounds the Juice of the Tabaco and the leaves stamped. And God would, that laying it upon the hurts, the griefs, madnes, and accidents wherewith they died were mittigated, and in such sorte they were delivered of that euill that the strength of the Venom was taken away and the wounds were healed, of the which there was great admiration. Which thing being knowen to them of

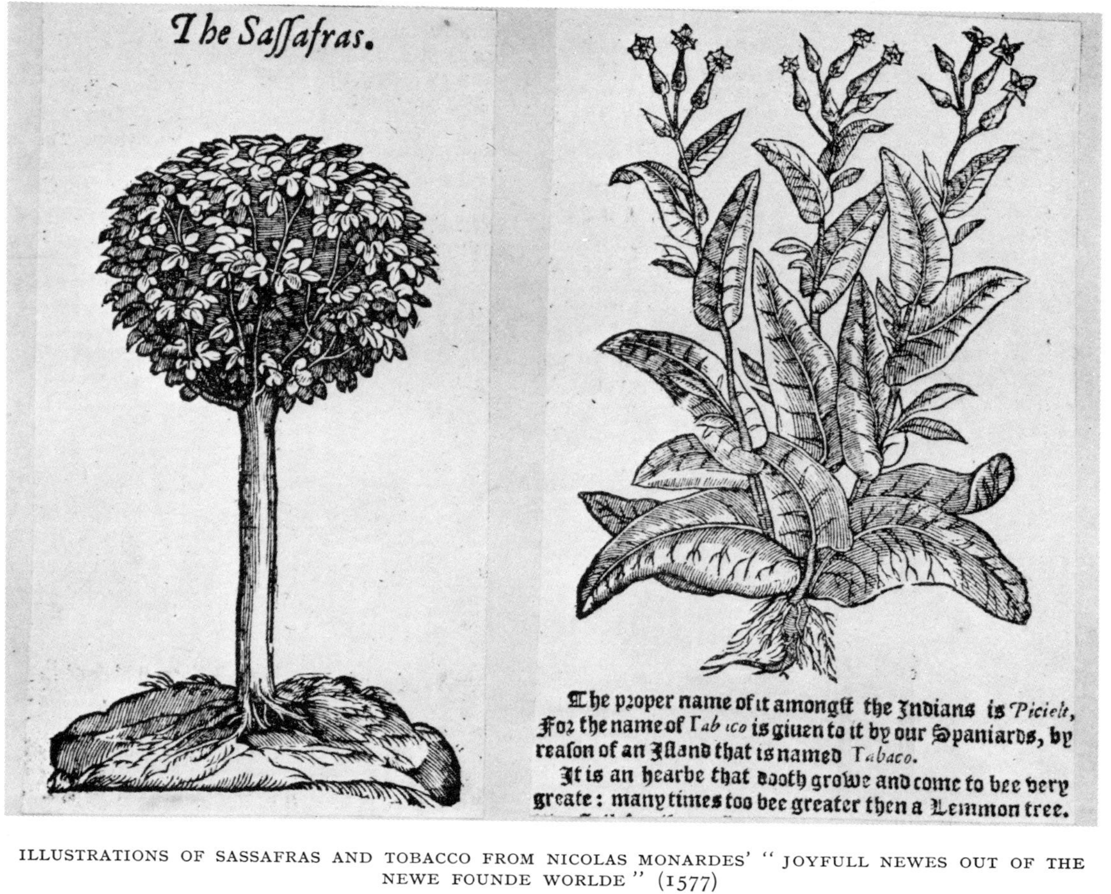

ILLUSTRATIONS OF SASSAFRAS AND TOBACCO FROM NICOLAS MONARDES' "JOYFULL NEWES OUT OF THE NEWE FOUNDE WORLDE" (1577)

(The figure of tobacco is the first printed illustration of that plant to appear in an English book)

the Islande they use it also in other hurtes and wounds, which they take when they fight with the wilde people: nowe they stand in no feare of them, by reason they have founde so great a remedie in a case so desperate. This Hearbe hath also vertue against the hearbe called of the Crosse boweshooter, which our hunters doe use to kill the wilde beastes withall and which hearbe is Venom most stronge, and doeth kill without remedie, which the Kinges pleasure was to prooue and commanded to make experience thereof, and they wounded a little dogge in the throate, and put forthwith into the wound the hearbe of the Crosse boweshooter, and after a little whyle, they powred into the self same wound that they had annointed with the Crosse boweshooters hearbe, a good quantitie of the juice of Tabaco and layde the stamped leaves upon it and they tied up the dogge and he escaped, not without great admiration of all men that saw him. Of the which the excellent Phisition of the Chamber of his Maiestie, Doctor Barnarde in the margent of this booke, that sawe it, by the commaundement of his Maiestie, writeth these wordes—'I made this experience by the commaundement of the Kinges Maiesty. I wounded the dogge with a knife and after I put the Crosse boweshooters hearbe into the wound and the hearbe was chosen and the dogge was taken of the hearbe, and the Tabaco and his Juyce being put into the wounde the dogge escaped and remained whole.'"

We are further given an exceptionally interesting account of the use of tobacco in the religious ceremonies of the Red Indians. "One of the meruelles of this hearbe and that whiche bringeth most admiration is the maner howe the Priests of the Indias did use it, which was in this maner: when there was amongst the Indians any maner of businesse of great importaunce, in the whiche the chiefe Gentleman called Casiques or any of the principall poople [people] of the Countrey had necessitie to consult with their Priestes in any businesse of importaunce: then they went and propounded their matter to their chiefe

Priest, foorthwith in their presence he tooke certeyne leaues of the Tabaco and cast them into ye fire and did receive the smoke of them at his mouth and at his nose with a Cane, and in taking of it he fell downe uppon the ground as a Dead man, and remayning so according to the quantity of the smoke that he had taken, when the hearbe had done his woorke he did revive and awake, and gave them then aunsweares [answers] according to the visions, and illusions whiche he sawe, whiles he was rapte in the same maner, and he did interprete to them as to him seemed best, or as the Divell had counselled him, giuing them continually doubtfull aunsweres in such sorte that howsoever it fell out, they might say that it was the same whiche was declared and the aunswere that he made.

"In like sort the rest of the Indians for their pastime do take the smoke of the Tabaco, to make themselves drunke withall, and to see the visions, and things that represent unto them, that wherein they do delight: and other times they take it to know their businesse and successe, because conformable to that whiche they haue seene, being drunke therewith, euen so they iudge of their businesse. And as the devil is a deceuer and hath the knowledge of the vertue of hearbs, so he did shew the vertue of this Hearb, that by the meanes thereof, they might see their imaginations and visions, that he hath represented unto them and by that meanes deceiue them."

The Red Indians also used this herb when they were obliged to travel for several days "in a dispeopled countrie where they shal finde neither water nor meate." They rolled the leaves into small balls, which they put "betweene the lower lippe and the teeth and goe chewing it all the time that they trauell and that whiche they chew they swallow downe and in this sort they journey three or foure dayes without hauing neede of meate or drink, for they feele no hunger nor weaknesse nor their trauel doth trouble them." (This custom Monardes compares to that of the bear, which during the winter "remaineth in his Caue and liueth without meate or drink, with onely chewing his pawes"!)

On its first introduction into Europe tobacco seems to have been regarded as a new all-heal, and in the city of Seville, we read, " they know not what other to doe, hauing cut or hurt themselves but to run to the Tabaco as to a most readie remedie. It doth meruellous workes, without any need of other Surgery, but this only hearbe." One chapter is devoted entirely to an account of various cures effected by tobacco, and it is interesting to read the authoritative account of the origin of the botanical name " Nicotiana." Monardes tells us that it was so called after Nicot, " my very friend ye first author inventer and bringer of this hearbe into France." It appears that " Maister John Nicot, being Embassador for his Maiestie in Portugall, in the yeere of our Lorde 1559, went one day to see the Prysons of the King of Portugall, and a Gentleman, being the Keeper of the said Prysons, presented him with this hearb as a strange plant brought from Florida." The same Maister Nicot, " hauing caused the said hearb to be set in his Garden, where it grewe and multiplyed maruellously," experimented with it, and amongst other things cured a young man who had a sore on his nose. Quite a number of cures were effected, the most interesting being that of one of Nicot's own cooks, who " hauing almost cutte off his thombe with a great Chopping Knife ran unto the said Nicotiane and healed it " !

The prescription for the ointment of tobacco is as follows :—" Take a pounde of the freshe leaues of the sayde Hearbe, stampe them, and mingle them with newe Waxe, Rosine, common oyle of each three ounces, let them boyle altogether, untill the Juice of Nicotiane be consumed, then add therto three ounces of Venise Turpentine, straine the same through a Linen cloth, and keepe it in Pottes to your use." The account of tobacco ends thus :—" Loe here you haue the true Historie of Nicotiane of the which the sayde Lorde Nicot, one of the Kinge's Counsellors, first founder out of this hearbe, hath made me privie, as well by woorde as by writing, to make thee (friendly Reader) partaker thereof, to whome I require thee to yeeld as harty thankes as I acknowledge myself bound unto him for this benefite received."

We find that the Indians first taught the Spaniards the use of sassafras, and " the Spaniards did begin to cure themselves with the water of this tree and it did in them greate effectes, that it is almost incredible: for with the naughtie meates and drinkyng of the rawe waters, and slepyng in the dewes, the moste parte of them came to fall into continuall Agues. . . . Thei tooke up the roote of this Tree and tooke a peece thereof suche as it seemed to theim beste, thei cutte it small into verie thinne and little peeces and cast them into water at discretion, little more or lesse, and thei sodde it the tyme that seemed nedefull for to remaine of a good colour, and so thei dranke it in the mornyng fastyng and in the daie tyme and at dinner and supper, without kepyng any more waight or measure, then I have saied, nor more keepyng, nor order then this, and of this thei were healed of so many griefes and euill diseases. That to heare of them what thei suffred and how thei were healed it doeth bryng admiration and thei whiche were whole dranke it in place of wine, for it doeth preserue them in healthe: As it did appeare verie well by theim, that hath come frō thence this yere, for thei came all whole and strong, and with good coulours, the whiche doeth not happen to them that dooeth come from those partes and from other conquestes, for thei come sicke and swolne, without collour, and in shorte space the moste of theim dieth: and these souldiours doeth trust so muche in this woodde that I beyng one daie amongest many of them, informing myself of the thynges of this Tree, the moste parte of them tooke out of their pokettes a good peece of this woodd, and said: 'Maister, doe you see here the woodde, that euery one of us doth bryng for to heale us with all, if we do fall sicke, as we haue been there,' and they began to praise so muche, to confirme the meruelous workes of it, with so many examples of them that were there, that surely I gave greate credite unto it and thei caused me to beleeve all that thereof I had heard, and gave me courage to experimente it as I have doen." There is another vivid glimpse of the use of sassafras as a pomander when the pestilence was rife in Seville. "Many did use to carrie a peece of the Roote of the wood with them to smell to it continually, as to a Pomander. For with his smell so acceptable it did rectifie the infected ayre: I caried with mee a peece a greate tyme, and to my seemyng I founde greate profite in it. For with it and with the chewing of the rinde of lemmon in the mornyng and in the daye tyme for to preserve health it hath a greate strength and property. It seemeth to mee that I was delivered by the healpe of God from the fyre in the whiche we that were Phisitions went in, blessed be our Lorde God that delivered us from so great euill and gave us this moste excellente Tree called Sassafras, which hath so greate vertues, and doth suche maruellous effectes as we have spoken of and more that the tyme will shewe us, which is the discouerer of all thinges."

It is a far cry from Monardes's book to that by "John Josselyn Gentleman," written nearly a hundred years later. Instead of the atmosphere of the El Dorado of the Spanish Main, of the galleons, of the tropical sun and plants of the West Indies, we find ourselves in the good company of the first settlers in New England, the Spanish Empire being only a memory of the past. Just fifty years after the landing of the Pilgrim Fathers on American soil, *New England's Rarities discovered* was printed at the Green Dragon in St. Paul's Churchyard, London, and the book is of peculiar interest, for it contains the first published lists of English plants that would thrive in America. There is a certain pathos in the efforts of the new settlers to produce in the New Country (which then took two months to reach) something that would remind them of the familiar English gardens of their old homes, and no one with a gardener's heart can read it without sympathy. The book was written by one John Josselyn, who undertook the then perilous voyage in order to stay with his only brother, who lived a hundred leagues from

Boston. There he remained about eight years, making it his business to collect all the information he could about plants that interested him. Even as late as 1663 the country was very imperfectly explored, for he gravely informs the reader that he cannot say whether New England is an island or not. He is not very sure whether even America is an island, but is confident that the Indians are closely allied to the Tartars.

But to turn to the subject-matter of the book. First we have a careful list of plants which the author found and which were common in England also, and—what is quite delightful—notes on the uses made of these plants by the Red Indians. For instance, they used white hellebore to cure their wounds, and John Josselyn tells us exactly how. They first rubbed racoon's grease or wild cat's grease on the wounds and then strewed the dried and powdered root on to it. They also applied the powdered root for toothache. Under the yellow-flowered water-lily we find a note to the effect that the Indians used the roots for food, and Josselyn seems to have tried them himself, for he says that they taste of sheep's liver. "The Moose Deer," he says, "feed much on them and the Indians choose this time when their heads are under water to kill them." From acorns the Indians made the oil with which they rubbed themselves. This was prepared by burning rotten maple wood to ashes and then boiling acorns with these ashes till the oil floated on the top. Of American walnuts and violets he had apparently a poor opinion, for he describes the walnuts as being not much bigger than a nutmeg and "but thinly replenished with kernels," and the violets as inferior to the English "Blew Violet." The most interesting of the recipes is that for the beer which he used to brew for Indians who came to him when they had bad colds. New Englanders who still possess treasured old housewives' books will probably find they have recipes for the same kind of beer; for it is typical of that commonly made in England in the seventeenth century and is strangely flavoured with elecampane, liquorice, sassafras, aniseed, and fennel seed. Then follows a list of plants peculiar to New England, with a long description of "Indian wheat," of which "the Flower [flour] makes excellent Puddens." Another plant described at length is the hollow-leaved lavender, but it is difficult to identify it from the illustration. The most interesting part of this list is that consisting of plants to which no English names had yet been given.

It is hard to believe that before the Pilgrim Fathers landed some of the commonest weeds were unknown in their new country. Yet we have John Josselyn's list of these, and it includes couch-grass, shepherd's purse, dandelion, groundsel, sow-thistle, stinging-nettle, mallows, plantain, wormwood, chickweed, mullein, knot-grass and comfrey. The plantain, one always learnt as a child, follows the English colonist wherever he goes, and there is curious confirmation in Josselyn's note that the Indians called this familiar weed "'Englishman's Foot,' as though it were produced by their treading." But the most fascinating list of all is that of the English garden-plants which those early settlers tried to grow, and it is impossible to read it without realising the loving care which must have been lavished on the southernwood, rosemary, lavender, and other plants imported from English gardens, which survived the long journey only to succumb to the rigours of the New England winter. There is something so naïve and appealing about this list, the first gardening link, as it were, between England and America, that I give it in full as it stands in the original :

Cabbidge growes there exceeding well
Lettice
Parsley, Marygold, French Mallowes, Chervil, Burnet,
Winter Savory, Summer Savory, Time, Sage, Carrots.
Parsnips of a prodigous size,
Red Beetes,
Radishes
Purslain

Pease of all sorts and the best in the world. I never heard of nor did see in Eight Years time one worm Eaten Pea.
Spearmint, Rew will hardly grow
Featherfew prospereth exceedingly.
Southernwood is no plant for this Country, Nor Rosemary, Nor
Bayes,
White Satten groweth pretty well, so doth
Lavender Cotton. But
Lavender is not for the Climate.
Penny Royal,
Smalledge
Ground Ivy or Ale Hoof.
Gillyflowers will continue Two Years.
Fennel must be taken up and kept in a Warm Cellar all the Winter.
Housleek prospereth notably,
Hollyhocks.
Enula Campana, in two Years time the Roots rot,
Comferie with white Flowers,
Coriander and
Dill and
Annis thrive exceedingly, but Annis Seed as also the Seed of Fennel seldom come to maturity; the Seed of Annis is commonly eaten of a fly.
Clary never lasts but one Summer, the
Roots rot with the Frost,
Sparagus thrives exceedingly so does
Garden Sorrel and
Sweet Bryer or Eglantine
Bloodwort but sorrily but
Patience and
English Roses very pleasantly.
Celandine by the West Country Men called Kenning Wort grows but slowly.

Muschata as well as in England.
Pepperwort flourisheth notably and so doth Tansie
Musk Mellons are better than our English and Cucumbers.
Pompions there be of several kinds; they are dryer than our English pompions and better tasted; You may eat them Green."

The book ends in a delightfully irrelevant fashion with a poem on an Indian squaw, introduced as follows:—" Now, gentle Reader, having trespassed upon your patience a long while in the perusing of these rude Observations, I shall, to make you amends, present you by way of Divertisement, or Recreation, with a Copy of Verses on the Indian Squa or Female Indian trick'd up in all her bravery."

The American Physitian; or a Treatise of the Roots, Plants, Trees, Shrubs, Fruit, Herbs, etc., growing in the English Plantations in America,[1] has, as its name implies, more of a medical character than the older books. In his preface the writer, William Hughes, tells us:—" 'Tis likely some may say need we trouble ourselves with those things we cannot reach? To such I answer, that the most part of them here mentioned which grow not in England already are brought over daily and made use of. . . . I suppose there are few but would gladly know that there are such things in the world, although scarcely any which care or desire to go to see them; I hope this Description which is as right to truth as I could possibly draw it, if my eyesight failed me not, may be acceptable, although it be far short of what I intended; it being my desire to have made it more compleat by one more voyage into those parts of the World, in which my endeavours should not have been found wanting for the bringing and fitting of Roots, Seeds and other Vegetables to our climate, for, to increase the number of Rarities which we have here in our Garden already; in the

[1] Published in London. See Bibliography, p. 217.

which I perceive much may be done, if further industry were used, but I have yet met with no opportunity to accomplish the same; and therefore hope that some others who have conveniency will do something herein for the promotion of further knowledge in these and many other excellent things which those parts afford, and we are yet unacquainted with. And whosoever is offended at this that I have here written, may let it alone; it forceth none to meddle with it: I know the best things displease some, neither was there ever any man yet that could please all people: but in hurting none, possibly I may please some; for whom only it is intended."

The book itself contains interesting accounts of yams, gourds, potatoes, prickly pears, maize (of turkeys fed on maize he says, "If I should tell how big some of their turkeys are I think I should hardly be believed"), cotton, pepper and sugar. His dissertation on the making of sugar is one of the earliest accounts of the process. Of the "Maucaw tree" he writes that "the seeds being fully ripe are of a pure crimson or reddish colour apt to dye the skin with a touch so that it cannot quickly be washed off." The Red Indians used these seeds to dye their skins, and Hughes remarks, "were some Ladies acquainted with this Rarity, doubtless they would give much for it." The longest section of the book deals with the cacao tree, its fruit and the making of chocolate. Cacao kernels were used as tokens and cacao plantations were entailed property. "In Carthagena, New Spain and other adjacent places, they do not only entail their Cacao Walks or Orchards on their Eldest Sons, as their Right of Inheritance (as Lands here in England are settled on the next Heir), but these cacao kernels have been, and are in so great esteem with them, that they pass between man and man for any merchandise, in buying and selling in the Markets, as the most current silver Coyn; as I have been told and as some credible Writers do affirm." There is a notable description of the making of chocolate by the servants "before they go forth to work in the Plantations in a morning and without which they are not well able to perform their most laborious employments in the Plantations, or work with any great courage until eleven a clock, their usual time of going to Dinner." A detailed account of the preparation of the drink ends with this vivid picture: "and then taking it off the Fire they pour it out of the Pot into some handsome large Dish or Bason: and after they have sweetened it a little with Sugar, being all together and sitting down round about it like good Fellows, everyone dips in his Calabash or some other Dish, supping it off very hot." He describes all sorts of ways of using the chocolate, the best in his opinion being that of the "Maroonoes Hunters and such as have occasion to travel the Country." They made it into "lozanges," which "exceed a Scotch-man's provision of Oatmeal and Water, as much (in my opinion) as the best Ox-beef for strong stomacks exceeds the meanest food." Chocolate, it will be remembered, became a very fashionable drink in England in the seventeenth century, but Hughes considers it inferior to the genuine stuff made in the Plantations. In fact, he cautions English people to procure their chocolate straight from Jamaica, and then to see themselves to the making of it according to his directions!

In spite of its impressive name, *The South-Sea Herbal containing the names, use, etc. of divers medicinal plants lately discovered by Pere L. Feuillee, one of the King of France's herbalists . . . much desired and very necessary to be known of all such as now traffick to the South-Seas or reside in those parts* (1715), is only eight pages long, five of which are devoted to figures of the plants. Nevertheless this now rare little pamphlet is valuable inasmuch as it is probably the first account in English of the medicinal plants of Peru and Chili. The writer—James Petiver—began life by serving his apprenticeship to Mr. Feltham, apothecary to St. Bartholomew's Hospital, London. He afterwards qualified as apothecary and became demonstrator of plants to the Society of Apothecaries. All his life he seems to have been rather a recluse, devoting his time to the study of natural history specimens sent him from all parts of the world. His herbarium, now in the Sloane Collection in the Victoria and Albert Museum, South Kensington, is exceptionally interesting, for Petiver appears to have had friends in all parts of the world, mostly sea-captains, who took delight in sending him treasures. The value of his collection may be judged from the fact that shortly before his death, Sir Hans Sloane offered him £4000 for it. His *South-Sea Herbal* is purely medicinal, except for an appeal to anyone living in Quito who " would be pleased to procure branches of the leaves of Jesuits' Bark or Quinquina with its Flowers and Fruit, which Favour should be acknowledged and more accurate Figures given of each if communicated to your humble servant." There is unfortunately now no copy extant of another of Petiver's pamphlets, *The Virtues of several Sovereign Plants found wild in Maryland with Remarks on them.* Apparently not many were printed, for there is a note to this effect at the end of the advertisement: " Divers of these Tracts are now so very scarce that of some of them there are not 20 left." Owing to the fact that nearly every page of illustrations in Petiver's works is dedicated to some friend who had sent him specimens, we have preserved for us the record of his numerous correspondents. These dedications are very pleasant reading:—

" To ye memory of y^t curious Naturalist and Learned Father, Geo Jos^ph Camel for many Observations and Things sent me."

" To ye memory of my curious Friend Mr. Sam Browne, Surgeon at Madrass, for divers Indian Plants, Shells, Seeds, etc."

" To Mr. George Bouchere, Surgeon, For divers Minorca Plants, Seed, etc."

" To Mr. Alexander Bartlet, Surgeon, For divers Cape and Moca Plants, Shells, etc."

" To Mr. George London, Late Gardiner to K. Will and Q. Mary."

" To ye memory of Mr. Willm Browne, Surgeon, who Presented me wth Divers Plants, Shells, etc."

" To His Hearty Friend, Mr. John Stocker, in gratitude for divers Plants, Shells, etc."

" To Mr. Claud Joseph, Geoffroy, Apothecary Chymist and Fellow of ye Academy Royall in Paris."

" To Mr. Charles Du-Bois, Treasurer of the East India Company."

" To the Honourable Dr. William Sherard, Consul of Smyrna."

" To Captain Jonathan Whicker for Divers Shells from St. Christophers."

" To his Curious Friend, Mr. John Smart, Surgeon, For Divers Plants, etc., from Hudson's Bay."

" To his kind Friend, Capt George Searle for divers Antego Shells, Coralls, etc."

" To Capt. Thomas Grigg at Antego in gratitude for divers Insects, Shells, etc."

" To that very obliging Gentlewoman, Madam Hannah Williams at Carolina."

CHAPTER VI

JOHN PARKINSON, THE LAST OF THE GREAT ENGLISH HERBALISTS

"For truly from all sorts of Herbes and Flowers we may draw matter at all times not only to magnifie the Creator that hath given them such diversities of formes sents and colours, that the most cunning Worke man cannot imitate, and such vertues and properties, that although wee know many, yet many more lye hidden and unknowne, but many good instructions also to ourselves. That as many herbes and flowers with their fragrant sweet smels doe comfort, and as it were revive the spirits and perfume a whole house: even so such men as live vertuously, labouring to doe good and profit the Church of God and the Commonwealth by their paines or penne, doe as it were send forth a pleasing savour of sweet instructions, not only to that time wherein they live, and are fresh, but being drye, withered and dead, cease not in all after ages to doe as much or more."—JOHN PARKINSON, *Paradisus*, 1629.

THE last of the great English herbalists was John Parkinson, the author of the famous *Paradisus* and also of the largest herbal in the English language, *Theatrum Botanicum*, which was published when the author was seventy-three. The latter was intended to be a complete account of medicinal plants and was the author's most important work, yet it is with the *Paradisus* (strictly not a herbal, but a gardening book), that his name is popularly associated. Of Parkinson himself we can learn very little. We know only that he was born in 1567, probably in Nottinghamshire, and that before 1616 he was practising as an apothecary and had a garden in Long Acre "well stored with rarities."[1] He was appointed Apothecary to James I., and after the publication of his *Paradisus* in 1629 Charles I. bestowed on him the title of Botanicus Regius Primarius. Amongst Parkinson's acquaintances mentioned in his books were the learned Thomas Johnson, who in 1633 emended and brought out a new edition of Gerard's

[1] See *Theatrum Botanicum*, p. 609.

Herball, John Tradescant,[1] the famous gardener, traveller and naturalist, and the celebrated physician, Sir Theodore Mayerne. Parkinson died in 1650 and was buried at St. Martin's in the Fields. There is a portrait of him in his sixty-second year prefixed to his *Paradisus,* and a small portrait by Marshall at the bottom of the title-page of his *Theatrum Botanicum.*

The full title of Parkinson's *Paradisus,* which in the dedicatory letter to Queen Henrietta Maria he truly describes as " this Speaking Garden," is inscribed on a shield at the bottom of the frontispiece. The first three words, " Paradisi in Sole," are a punning translation into Latin of his own surname.

At the top of the page is the Eye of Providence with a Hebrew inscription, and on each side a cherub symbolising the winds. In the centre is a representation of Paradise with Adam grafting an apple tree and Eve running downhill to pick up a pineapple. The flowers depicted are curiously out of proportion, for the tulip flower is a good deal larger than Eve's head, and cyclamen in Paradise seems to have grown to a height of at least five feet.

The most interesting feature of this elaborately illustrated title-page is the representation of the " Vegetable Lamb " growing on a

[1] Both John Tradescant and his son were gardeners to Charles I. and Henrietta Maria. John Tradescant the elder is said by Anthony à Wood to have been a Fleming or a Dutchman, but this is doubtful. The name is neither Flemish nor Dutch but probably English, and in the inscription on his tomb in Lambeth Churchyard he and his son are described as " both gardeners to the rose and lily queen." This was Henrietta Maria. Parkinson in his *Paradisus* speaks of him as " that painfull industrious searcher and lover of all nature's varieties." Tradescant accompanied Sir Dudley Digges on his voyage round the North Cape to Archangel, and on his return wrote an account of the plants he had found in Russia—the earliest extant record of plants in that part. It is interesting to note that in this he compares the soil of Russia to that of Norfolk. In 1620 Tradescant joined an expedition against the Algerine corsairs as a gentleman volunteer, and he also accompanied the Duke of Buckingham (George Villiers), to whom he had formerly been gardener, on the ill-fated expedition to La Rochelle. On Buckingham's death he entered the royal service, and probably at this time established his well-known physic garden and museum at Lambeth. The house was called Tradescant's ark. There are three unsigned and undated portraits of the elder Tradescant in the Ashmolean Collection at Oxford.

stalk and browsing on the herbage round about it.[1] This records one of the most curious myths of the Middle Ages. The creature was also known as the Scythian Lamb and the Borametz or Barometz, a name derived from a Tartar word signifying "lamb." It was supposed to be at once a true animal and a living plant, and was said to grow in the territory of the "Tartars of the East," formerly called Scythia. According to some writers, the lamb was the fruit of a tree, whose fruit or seed-pod, when fully ripe, burst open and disclosed a little lamb perfect in every way. This was the subject of the illustration, "The Vegetable Lamb plant," in Sir John Mandeville's book. Other writers described the lamb as being supported above the ground by a stalk flexible enough to allow the animal to feed on the herbage growing near. When it had consumed all within its reach the stem withered and the lamb died. This is the version illustrated on Parkinson's title-page. It was further reported that the lamb was a favourite food of wolves, but that no other carnivorous animals would attack it. This remarkable legend obtained credence for at least 400 years. So far as is known, the first mention of it in an English book is the account given by Sir John Mandeville, "the Knyght of Ingelond that was y bore in the toun of Seynt Albans, and travelide aboute in the worlde in many diverse countries to se mervailes and customes of countreis and diversiters of folkys and diverse shap of men and of beistis." It is in the chapter describing the curiosities he met with in the dominions of the "Cham" of Tartary that the passage about the vegetable lamb occurs.[2] The origin of this extraordinary myth is undoubtedly

[1] It also figures on the title-page of Parkinson's *Theatrum Botanicum.*

[2] "Now schalle I seye you semyingly of Countries and Yles that bea beyonde the Countries that I have spoken of. Wherefore I seye you in pessynge be [by] the Lord of Cathaye toward the high Ynde and towards Bacharye, men passen be a Kyngdom that men clepen Caldhille, that is a fair contree. And there growethe a maner of Fruyt, as though it weren Gowrdes, and when thei ben rype men kutten hem ato, and men fynden with inne a lytylle Best, in Flesche, in Bon and Blode, as though it were a lytylle Lamb withouten wolle. And men eten both the Frut and the Best, and that is a great Marveylle. Of that Frute I have eaten, alle thoghe it were wonderfulle but that I knowe wel that God is marveyllous in his Werkes."

PARADISI IN SOLE
Paradiſus Terreſtris.
or
A Garden of all ſorts of pleaſant flowers which our English ayre will permitt to be nourſed vp:
with
A Kitchen garden of all manner of herbes, rootes, & fruites, for meate or ſawſe vſed with vs,
and
An Orchard of all ſorte of fruitbearing Trees and ſhrubbes fit for our Land
together
With the right orderinge planting & preſeruing of them and their vſes & vertues
Collected by John Parkinson
Apothecary of London.
1629

Qui veut parangonner l'artifice a Nature
Et nos Peres a l'Eden indiscret il mesure.

Le pas de l'Elephant par le pas du ciron,
Et de l'Aigle le vol pareil du moucheron.

TITLE-PAGE OF PARKINSON'S "PARADISUS" (1629)

to be found in the ancient descriptions of the cotton plant by Herodotus, Ctesias, Strabo, Pliny and others.[1] The following passages in Herodotus and Pliny will suffice to show how easily the myth may have grown. "Certain trees bear for their fruit fleeces surpassing those of sheep in beauty and excellence" (Herodotus). "These trees bear gourds the size of a quince which burst when ripe and display balls of wool out of which the inhabitants make cloths like valuable linen" (Pliny).

In his *Theatrum Botanicum* Parkinson describes the "Scythian Lamb," and one gathers that he accepted the travellers' tales about it. "This strange living plant as it is reported by divers good authors groweth among the Tartares about Samarkand and the parts thereabouts rising from a seede somewhat bigger and rounder than a Melon seede with a stalk about five palmes high without any leafe thereon but onely bearing a certaine fruit and the toppe in forme resembling a small lambe, whose coate or rinde is woolly like unto a Lambe's skinne, the pulp or meat underneath, which is like the flesh of a Lobster, having it is sayed blood also in it; it hath the forme of an head hanging down and feeding on the grasse round about it untill it hath consumed it and then dyeth or else will perish if the grasse round about it bee cut away of purpose. It hath foure legges also hanging downe. The wolves much affect to feed on them."

The preface to the *Paradisus* is singularly beautiful, being typical of the simple, devout-minded author, but it is too long to quote. The book itself is truly "a speaking garden," a tranquil, spacious Elizabethan garden, full of the loveliness, colour and scent of damask, musk and many other roses; of lilies innumerable—the crown imperial, the gold and red lilies, the Persian lily ("brought unto Constantinople and from thence sent unto us by Mr. Nicholas Lete, a worthy Merchant and a lover of all faire flowers"), the blush Martagon, the bright red Martagon of

[1] See Herodotus (lib. iii. cap. 106), Ctesias (*Indica*); Strabo (lib. xv. cap. 21); Theophrastus *De Historia Plantarum* (lib. iv. cap. 4); Pliny, *Naturalis Historia*.

Hungary and the lesser mountain lily. Of fritillaries of every sort—of which Parkinson tells us that " although divers learned men do by the name given unto this delightful plant think it doth in some things partake with a Tulipe or Daffodill; yet I, finding it most like unto a little Lilly, have (as you see here) placed it next unto the Lillies and before them." Of gay tulips, which were amongst his special favourites—" But indeed this flower, above many other, deserveth his true commendations and acceptance with all lovers of these beauties, both for the stately aspect and for the admirable varietie of colour, that daily doe arise in them,"—and of which he had a collection such as would be the glory of any garden—the tulip of Caffa, the greater red Bolonia tulip, the tulip of Candie, the tulip of Armenia, the Fool's Coat tulip, the Cloth of Silver tulip and others too numerous to mention. (" They are all now made denizens in our Gardens," he joyously tells us, " where they yield us more delight and more increase for their proportion by reason of their culture, than they did unto their owne naturals "). Of daffodils, crocuses and hyacinths in boundless profusion, amongst which are to be noted many pleasing names that we no longer use. Of asphodels, " which doe grow naturally in Spaine and France and from thence were first brought unto us to furnish our Gardens." Of many-coloured flags, which he calls by the prettier name of " flower de luce," and amongst which he gives pride of place " for his excellent beautie and raretie to the great Turkie Flower de luce." Of gladioli, cyclamen and anemones. Of the last-named he writes thus :—

" The Anemones likewise or Windeflowers are so full of variety and so dainty so pleasant and so delightsome flowers that the sight of them doth enforce an earnest longing desire in the mind of anyone to be a possessoure of some of them at the leaste. For without all doubt this one kind of flower, so variable in colours, so differing in form (being almost as many sortes of them double as single), so plentifull in bearing flowers and so durable in lasting and also so easie both to preserve and to encrease is of itselfe alone almost sufficient to furnish a garden with flowers for almost half the yeare. But to describe the infinite (as I may so say) variety of the colours of the flowers and to give each his true distinction and denomination it passeth my ability I confesse, and I thinke would grauell the best experienced in Europe." (Nevertheless he writes of about fifty varieties.) Of fragrant crane's-bills, bear's-ears, primroses and cowslips. Of violets, borage, marigolds, campions, snapdragons, columbines and lark's-heels (delphiniums). Of gillyflowers (why have we given up this old-fashioned English name?), and how pleasant is the mere reading of his list of varieties—" Master Bradshawe his daintie Ladie," "Ruffling Robin," "The Fragrant," "The Red Hulo," "John Witte his great tawny gillow flower," "Lustie Gallant," "The fair maid of Kent," "The Speckled Tawny." "But the most beautiful that ever I did see was with Master Ralph Tuggie,[1] the which gilliflower I must needes therefore call 'Master Tuggies Princesse,' which is the greatest and fairest of all these sorts of variable tawnies, being as large fully as the Prince or Chrystall, or something greater, standing comely and round, not loose or shaken, or breaking the pod as some other sorts will; the marking of the flower is in this manner: It is of a stamell colour, striped and marbled with white stripes and veines quite through every leafe, which are as deeply iagged as the Hulo: sometimes it hath more red then white, and sometimes more white then red, and sometimes so equally marked that you cannot discern which hath the mastery; yet which of these hath the predominance, still the flower is very beautifull and exceeding delightsome." Of peonies, lupins, pinks, sea-holly and sweet-william. Of lilies of the valley, gentian, Canterbury bells, hollyhocks and mallows ("which for their bravery are entertained everywhere unto every countrey-woman's garden"). Of fox-gloves, goldilocks, valerian and mullein. Of cuckoo-flowers, "or Ladies smockes," both the double and the trefoil. The first

1 "Master Tuggie," who lived in Westminster, was a famous grower of gilliflowers. See p. 116.

kind, Parkinson tells us, "is found in divers places of our owne Countrey as neere Micham about eight miles from London;" also in Lancashire, "from whence I received a plant, which perished, but was found by the industrie of a worthy Gentlewoman dwelling in those parts called Mistresse Thomasin Tunstall, a great lover of these delights. The other was sent me by my especiall good friend John Tradescant, who brought it among other dainty plants from beyond the seas, and imparted thereof a root to me." Of clematis and candytufts, honeysuckles and jasmine. Of double-flowered cherries, apples and peaches. "The beautiful shew of these three sorts of flowers," he says, "hath made me to insert them into this garden, in that for their worthinesse I am unwilling to bee without them, although the rest of their kindes I have transferred into the Orchard, where among other fruit trees they shall be remembered: for all these here set downe seldome or never beare any fruite, and therefore more fit for a Garden of flowers then an Orchard of fruite. These trees be very fit to be set by Arbours."

In this garden of pleasant flowers we find also many fragrant herbs. "After all these faire and sweete flowers," says Parkinson, "I must adde a few sweete herbes, both to accomplish this Garden, and to please your senses, by placing them in your Nosegayes, or elsewhere as you list. And although I bring them in the end or last place, yet they are not of the least account." He writes first of rosemary, the common, the gilded, the broad-leaved and the double-flowered. Of rosemary he tells us: "This common Rosemary is so well knowne through all our Land, being in every woman's garden, that it were sufficient but to name it as an ornament among other sweete herbes and flowers in our Garden. It is well observed, as well in this our Land (where it hath been planted in Noblemen's, and great men's gardens against brick wals, and there continued long) as beyond the Seas, in the naturall places where it groweth, that it riseth up in time unto a very great height, with a great and woody stemme (of that compasse that—being clouen out into thin boards—it hath served to make lutes, or such like instruments, and here with us Carpenters rules, and to divers other purposes), branching out into divers and sundry armes that extend a great way, and from them againe into many other smaller branches, whereon we see at several distances, at the ioynts, many very narrow long leaves, greene above, and whitish underneath, among which come forth towards the toppes of the stalkes, divers sweet gaping flowers of a pale or bleake blewish colour, many set together standing in whitish huskes . . . although it will spring of the seede reasonable well, yet it is so small and tender the first yeare, that a sharpe winter killeth it quickly, unlesse it be very well defended; the whole plant as well leaves as flowers, smelleth exceeding sweete." Of sage and of lavender both the purple and the rare white [1] ("there is a kinde hereof that beareth white flowers and somewhat broader leaves, but it is very rare and seene but in few places with us, because it is more tender, and will not so well endure our cold Winters"). "Lavender," he says, "is almost wholly spent with us, for to perfume linnen, apparell, gloues and leather and the dryed flowers to comfort and dry up the moisture of a cold braine." Of French lavender ("the whole plant is somewhat sweete, but nothing so much as Lavender). It groweth in the Islands Staechades which are over against Marselles and in Arabia also: we keep it with great care in our Gardens. It flowreth the next yeare after it is sowne, in the end of May, which is a moneth before any Lavender." Of lavender cotton, of which he writes: "the whole plant is of a strong sweete sent, but not unpleasant, and is planted in Gardens to border knots with, for which it will abide to be cut into what forme you think best, for it groweth thicke and bushy, very fit for such workes, besides the comely shew the plant it selfe thus wrought doth yeeld, being alwayes greene and of a sweet sent." Of basil, "wholly spent to make sweet or washing waters, among other sweet herbes, yet sometimes it is put into nosegayes. The Physicall properties are to procure a cheerfull and merry heart"; and marjoram, "not onely much used to please the outward

[1] White lavender was a favourite with Queen Henrietta Maria.

senses in nosegayes and in the windowes of houses, as also in sweete pouders, sweete bags, and sweete washing waters." Of all the varieties of thyme and hyssop—and of the white hyssop he writes that its striped leaves "make it delightfull to most Gentlewomen." Hyssop, he tells us further, "is used of many people in the Country to be laid unto cuts or fresh wounds, being bruised, and applyed eyther alone, or with a little sugar." "And thus," he concludes this part of the book, "have I led you through all my Garden of Pleasure, and shewed you all the varieties of nature housed therein, pointing unto them and describing them one after another. And now lastly (according to the use of our old ancient Fathers) I bring you to rest on the Grasse, which yet shall not be without some delight, and that not the least of all the rest."

From his garden of pleasant flowers he leads us to the kitchen garden, full not only of "vegetables" as we understand the term, of strawberries, cucumbers and pompions, but also of a vast number of herbs in daily use, many of them never seen in modern gardens. Besides the familiar thyme, balm, savory, mint, marjoram, and parsley, there are clary, costmary, pennyroyal, fennel, borage, bugloss, tansy, burnet, blessed thistle, marigolds, arrach, rue, patience, angelica, chives, sorrel, smallage, bloodwort, dill, chervil, succory, purslane, tarragon, rocket, mustard, skirrets, rampion, liquorice and caraway. But according to Parkinson they used fewer herbs in his day than in olden times; for under pennyroyal we find, "The former age of our great-grandfathers had all these pot herbes in much and familiar use, both for their meates and medicines, and therewith preserved themselves in long life and much health: but this delicate age of ours, which is not pleased with anything almost, be it meat or medicine, that is not pleasant to the palate, doth wholly refuse these almost, and therefore cannot be partaker of the benefit of them." From the kitchen garden with all these herbs, "of most necessary uses for the Country Gentlewomen's houses," he leads us, finally, to the orchard, with its endless varieties of apple and pear trees, of cherries, medlars, plums, "apricockes" and nectarines, of figs and peaches and almonds, of quinces, walnuts, mulberries and vines (ending with the Virginian vine, of which he says, "we know of no use but to furnish a Garden and to encrease the number of rarities"), until, like the Queen of Sheba, we feel that, with all we have heard of the comfortable splendour of Elizabeth's reign, the half has not been told us. "And thus," Parkinson concludes, "have I finished this worke, and furnished it with whatsoever Art and Nature concurring could effect to bring delight to those that live in our Climate and take pleasure in such things; which how well or ill done, I must abide every one's censure; the iudicious and courteous I onely respect, let Momus bite his lips and eate his heart; and so Farewell."

Parkinson's monumental work, *Theatrum Botanicum*, was completed, as already mentioned, in his seventy-third year. In it about 3800 plants are described (nearly double the number of those in the first edition of Gerard's Herbal). In the *Theatrum* he incorporated nearly the whole of Bauhin's *Pinax*, besides part of the unfinished work by de l'Obel mentioned before. The book remained the most complete English treatise on plants until the time of Ray. Parkinson originally intended to entitle it "A Garden of Simples"[1] and, had he done so, it is at least possible that this work, to which he devoted the greater part of his life, would have achieved the popularity it deserved. Except in the illustrations, it is a finer book than Gerard's, but the latter remained the more popular. In fact, this herbal of Parkinson's is an outstanding proof that a good book may be ruined by a bad title. *Theatrum Botanicum* sounds hard and chilling, whereas *Gerard's Herball* has an attractive ring. The fact that the former never attained the popularity achieved by the latter seems the more pathetic when we read the author's

[1] This he tells us at the end of the preface to the *Paradisus*. "Thus have I shewed you both the occasion and scope of this Worke, and herein have spent my time, paines, and charge, which if well accepted, I shall thinke well employed, and may the sooner hasten the fourth Part, A Garden of Simples; which will be quiet no longer at home, then that it can bring his Master newes of faire weather for the iourney."

own concluding charge to this work of his lifetime :—"Goe forth now therefore thou issue artificial of mine and supply the defect of a Naturall, to beare up thy Father's name and memory to succeeding ages and what in thee lyeth effect more good to thy Prince and Country then numerous of others, which often prove rather plagues then profits thereto, and feare not the face of thy fiercest foe."

The ornamental title-page of the *Theatrum Botanicum* is both interesting and impressive. The two most important figures are those of Adam and Solomon (representing Toil and Wisdom respectively). Solomon is dressed in a long coat with an ermine cape, and he wears Roman sandals. At the four corners of the page are female figures :—Europe driving majestically in a chariot with a pair of horses; Asia clad in short skirts and shoes with curled points and riding a rhinoceros; Africa wearing only a hat, and mounted upon a zebra; and America, also unclothed, carrying a bow and arrow and riding a sheep with surprisingly long ears. Each of these figures is surrounded by specimens of the vegetation of their respective continents.

It is curious to find in the dedicatory letter to Charles I a touch of the old belief that diseases are due to evil spirits :—

"And I doubt not of your Majesties further care of their bodies health that such Workes as deliver approved Remedyes may be divulged whereby they may both cure and prevent their diseases. Most properly therefore doth this Worke belong to your Majesty's patronage both to further and defend that malevolent spirits should not dare to cast forth their venome or aspertions to the prejudice of any well-deserving, but that thereby under God and Good direction, all may live in health as well as wealth, peace and godliness, which God grant and that this boldnesse may be pardoned to

"Your Majestyes
"Loyale Subject
"Servant and Herbarist
"JOHN PARKINSON."

TITLE-PAGE OF PARKINSON'S "THEATRUM BOTANICUM" (1640)

There are letters extolling the Herbal from three Oxford doctors, two of whom refer to the then newly-made physic garden on the Cherwell. One writes thus: " Oxford and England are happy in the foundation of a spacious illustrious physicke garden, compleately beautifully walled and gated, now in levelling and planting with the charges and expences of thousands by the many wayes Honourable Earle of Danby, the furnishing and enriching whereof and of many a glorious Tempe, with all usefull and delightfull plants will be the better expedited by your painefull happy satisfying Worke.

"Tho. Clayton, His Majesty's prof. of Physicke, Oxon."

One who signs himself " Your affectionate friend John Bainbridge Doctor of Physique, and Professor of Astronomy, Oxon " writes thus: " I am a stranger to your selfe but not to your learned and elaborate volumnes. I have with delight and admiration surveyed your *Theatrum Botanicum*, a stately Fabrique, collected and composed with excessive paines. . . . It is a curious pourtrait and description of th' Earths flowred mantle, the Herbarist's Oracle, a rich Magazin of soveraigne Medicines, physicall experiments and other rarities."

Parkinson divides his plants into " Classes or Tribes " :—

1. Sweete smelling Plants.
2. Purging Plants.
3. Venemous Sleepy and Hurtfull plants and their Counter poysons.
4. Saxifrages.
5. Vulnerary or Wound Herbs.
6. Cooling and Succory like Herbs.
7. Hot and sharpe biting Plants.
8. Umbelliferous Plants.
9. Thistles and Thorny Plants.
10. Fearnes and Capillary Herbes.
11. Pulses.
12. Cornes.

13. Grasses Rushes and Reeds.
14. Marsh Water and Sea plants and Mosses and Mushromes.
15. The Unordered Tribe.
16. Trees and Shrubbes.
17. Strange and Outlandish Plants.

Under " The Unordered Tribe " we find the naïve remark: " In this tribe as in a gathering campe I must take up all those straglers that have either lost their rankes or were not placed in some of the foregoing orders that so I may preserve them from losse and apply them to some convenient service for the worke " !

It is surprising how much folk lore survives even in Parkinson's Herbal. Like Gerard, he pours scorn on a good many contemporary beliefs, but many he accepts unquestioningly, especially those concerning the use of herbs as amulets and also for the promotion of happiness. He gives also some old gardening beliefs not to be found in other herbals, but very common in contemporary books on gardening and husbandry, and more bee lore than most herbals contain. Nearly all the old herbalists believed in the value of growing balm near the beehives, and also of rubbing the hive with this herb, but Parkinson alone tells us of the harmful effects of woad: [1] " Some have sowen it but they have founde it to be the cause of the Destruction of their Bees, for it hath been observed that they have dyed as it were of a Flix that have tasted hereof." Of balm,[2] however, he writes: " it is an hearbe wherein Bees do much delight both to have their Hives rubbed therewith to keepe them together and draw others and for them to suck and feed upon." Elsewhere he tells us that " it hath been observed that bees will hardly thrive well where many Elmes doe grow or at least if they upon their first going forth abroad after Winter doe light on the bloomings or seed thereof." [3] Of the sweet-smelling flag he says: " it is verily believed of many that the leaves

[1] *Theatrum Botanicum*, p. 601.
[2] *Ibid.*, p. 43.
[3] *Ibid.*, p. 1405.

or roots of Acorus tyed to a hive of Bees stayeth them from wandering or flying away and draweth a greater resort of others thereto." [1]

Upon the use of herbs as amulets his views seem inconsistent. He is scornful of the custom of hanging a piece of mistletoe to children's necks "against witchcraft and the illusion of Sathan"; yet he gravely informs us that "if the sope that is made of the lye of the ashes [of glassewort] be spread upon a piece of thicke course brown paper cut into the forme of their shooe sole, that are casually taken speechless and bound to the soles of their feete it will bring again the speech and that within a little time after the applying thereof if there be any hope of being restored while they live: this hath been tried to be effectuall upon diverse persons.[2] The custom of wearing meadowsweet or hanging it up in living-rooms [3] he describes as a "superstitious conceit," but he accepts without demur the tradition [4] that a wreath of periwinkle "worne about the legs defendeth them that wear it from the crampe." Bartholomæus Anglicus tells us that Augustus Cæsar used to wear a wreath of bryony during a thunderstorm to protect himself from lightning, but the story is not repeated until, after the lapse of four hundred years, we find in Parkinson the statement that "Augustus Cæsar was wont to weare bryony with bayes made into a roule or garlande thereby to be secured from lightening." [5] Parkinson regards the use of herbs against witchcraft as sheer foolishness, but he is the only herbalist who gives us a potion [6] which "resisteth such charmes or the like witchery that is used in such drinkes that are given to produce love." Like Gerard, he does not question the efficacy of borage, bugloss and

[1] *Theatrum Botanicum.*, p. 144. [2] *Ibid.*, p. 281.
[3] *Ibid.*, p. 265. [4] *Ibid.*, p. 384. [5] *Ibid.*, p. 181.
[6] *Ibid.*, p. 422. Of this "Indian Spanish Counter poyson" Parkinson gives us the further interesting information that "the Indians doe not eate the bodies of those they have slaine by their poysoned arrowes untill they have lyen three or foure dayes with their wounds washed with the juice of this herbe; which rendereth them tender and fit to be eaten which before were hard."

many other herbs to promote happiness. Of borage [1] he tells us: "The leaves floures and seedes are very cordiall and helpe to expell pensivenesse and melancholie that ariseth without manifest cause"; and of a confection made from oak galls,[2] that it is "dayly commended and used with good effect against Melancholy passions and sorrow proceeding of no evident cause." Water yarrow "is taken with vinegar to helpe casuall sighings also the Toothache." [3] Under viper's-grass [4] we find "the water distilled in glasses or the roote itself taken is good against the passions and tremblings of the heart as also against swoonings sadnes and melancholy," and under bugloss,[5] that "the rootes or seedes are effectuall to comfort the heart and to expell sadnesse and causelesse melancholy." In common with other herbalists he believed also that herbs could be used to strengthen the memory, to help weak brains, to quicken the senses and even to soothe "frenzied" people. Of eyebright,[6] used for so many centuries, and even until recent times, to help dull sight, he says: "it helpeth a weake braine or memory and restoreth them being decayed in a short time." Fleabane "bound to the forehead is a great helpe to cure one of the frensie," while "the distilled water of thyme applyed with vinegar of Roses to the forehead easeth the rage of Frensye." [7] Lavender is of "especiall good use for all griefes and paines of the head and brain," [8] and sage [9] is of "excellent good use to helpe the memory by warming and quickening the senses."

Parkinson gives more beauty recipes than any other herbalist. For those who wish to darken their hair he recommends washing it with a decoction of bramble leaves.[10] The golden flowers of mullein [11] "boyled in lye dyeth the haires of the head yellow and maketh them faire and smooth." The ashes of southernwood [12] mixed with old salad oil will cause a

[1] *Theatrum Botanicum*, p. 767. [2] *Ibid.*, p. 1397.
[3] *Ibid.*, p. 1259. [4] *Ibid.*, p. 410. [5] *Ibid.*, p. 518.
[6] *Ibid.*, p. 1330. [7] *Ibid.*, p. 128. [8] *Ibid.*, p. 74.
[9] *Ibid.*, p. 54. [10] *Ibid.*, p. 1016. [11] *Ibid.*, p. 63.
[12] *Ibid.*, p. 95.

beard to grow or hair on a bald head, and yarrow is almost as good; garden spurge, elder flowers, broom, madder, rue, gentian, scabious, betony, elecampane, Solomon's Seal, the great hawkweed and lupin are all excellent to "cleanse the skinne from freckles, sunburn and wrinkles."[1] The French women "account the distilled water of pimpernell mervailous good to clense the skinne from any roughnesse deformity or discolouring thereof and to make it smooth neate and cleere."[2] The Italian dames, however, "doe much use the distilled water of the whole plant of Solomon's Seal."[3] Lupin seems to have the most remarkable virtue, for not only will it take away all smallpox marks, but it will also make the user "look more amiable"! Many women, therefore, "doe use the meale of Lupines mingled with the gall of a goate and some juyce of Lemons to make into a forme of a soft ointment."[4] Parkinson is the only herbalist who gives recipes to enable people to get thin and also to look pale. "The powder of the seedes of elder[5] first prepared in vinegar and then taken in wine halfe a dramme at a time for certaine dayes together is a meane to abate and consume the fat flesh of a corpulent body and to keepe it leane." For those who like to look pale he recommends cumin seed and bishopsweed.[6] And "for a sweet powder[7] to lay among linnen and garments and to make sweet waters to wash hand-gloves or other things to perfume them" he recommends the roots of the sweet-smelling flag.

It is, however, the curious out-of-the-way pieces of information on all sorts of matters which are so interesting in Parkinson's Herbal. He tells us that three several sorts of colours are made from the berries of the purging thorn; that the yellow dye is used by painters, "and also by Bookbinders to colour the edges of Bookes and by leather dressers to colour leather"; that the green dye is "usually put up into great bladders tyed

[1] *Theatrum Botanicum*, pp. 135, 191, 210, 233, 275, 408, 492, 613, 652, 693, 700, 790, 1075. [2] *Ibid.*, p. 559. [3] *Ibid.*, p. 700. [4] *Ibid.*, p. 1075. [5] *Ibid.*, p. 210. [6] *Ibid.*, pp. 888 and 913. [7] *Ibid.*, p. 144.

with strong thred at the head and hung up untill it is drye, which is dissolved in water or wine, but sacke is the best to preserve the colour from 'starving,' as they call it, that is from decaying, and to make it hold fresh the longer"; and that the purple dye is made by leaving the berries on the bushes until the end of November, when they are ready to drop off. That the best mushrooms grow under oaks or fir trees. That spurry leaves bruised and laid to a cut finger will speedily heal it, "whereof the Country people in divers places say they have had good experience," and that it is also good for causing "the Kine to give more store of milke than ordinary otherwise, so it causeth Pullaine likewise to lay more store of egges." That the fruit of the bead tree "being drilled and drawne on stringes serves people beyond sea to number their prayers thereon least they forget themselves and give God too many." That in Warwickshire the female fern was always used "in steed of Sope to wash their clothes," and that it was gathered about Midsummer, "unto good big balls which when they will use them they burne them in the fire until it becomes blewish, which being then layd by will dissolve into powder of itselfe, like unto Lime: foure of these balles being dissolved in warme water is sufficient to wash a whole bucke full of clothes." That the burning of lupin seeds drives away gnats, and that half-sodden barley "given to Hennes that hardly or seldome lay egges will cause them to lay both greater and more often." That country housewives use that common weed horsetail to scour their wooden, pewter and brass vessels, and sometimes boil the young tops of the same weed and eat them like asparagus. That bramble leaves do not fall until all the sharp frosts are over, "whereby the country men do observe that the extremity of Winter is past when they fall off." That every year sacks full of violets are sent from Marseilles to Alexandria and other parts of Egypt, "where they use them boyled in water which only by their religion they are enjoined to drinke." That if you suspect your wine is watered "you shall put some thereof into a cup that is made of ivie wood and if there be any water therein it will remaine in the cup and the wine will soak through, for the nature of Ivie is not to hold any wine so great an antipathy there is between them." That skilful shepherds are careful not to let their flocks feed in pastures where mouseare abounds, "lest they grow sicke and leane and die quickly after." That writing-ink can be made of the green fruit of alder trees. That the bark of the same tree is useful for making "a blacke dye for the courser sorts of things," and that the leaves put under the bare feet of travellers are "a great refreshing unto them." That the rose of Jericho opened the night our Saviour was born, and that placed in any house it will open when a child is born. That mouseare if given to any horse "will cause that he shall not be hurt by the Smith that shooeth him." That purslane is not only a sovereign remedy for crick in the neck, but also for "blastings by lightening, or planets and for burnings by Gunpowder or otherwise." That country folk in Kent and Sussex call sopewort "Gill-run-by-the-streete." That agrimony leaves will cure cattle suffering from coughs, and that wounded deer use this same herb to heal their hurts. That a decoction made of hemp will draw earthworms out of their holes and that fishermen thus obtain their bait. That crops of woad may be cut three times in the year, and that dyers' weed will change to green any cloth or silk first dyed blue with woad, "and for these uses there is great store of this herbe spent in all countries and thereof many fields are sowen for the purpose." That country-folk use goose-grass as a strainer "to clear their milke from strawes, haires, and any other thing that falleth into it." That St. John's wort is used by country-folk to drive away devils. That "Clownes woundwort" owes its name to a labourer who healed himself therewith of a cut with a scythe in his leg. That willow-herb, being burned, "driveth away flies and gnats and other such like small creatures which use in diverse places that are neere to Fennes, marsh or water sides to infest them that dwell there in the night season to sting and bite them, leaving the marks and spots thereof in their faces which beside the deformity, which is but for a while, leaveth them that are thus bitten not without paine for a time." That from turnesole (heliotropium) are made "those ragges of cloth which are usually called Turnesole in the Druggists and Grocers shoppes and with all other people and serveth to colour jellies or other things as every one please." That when French ladies coloured their faces with an ointment containing anchusa the colour did not last long. That no "good gentlewoman in the land that would do good" should be without a store of bugloss ointment either for her own family "or other her poor neighbours that want helpe and means to procure it," and that beyond the sea in France and Germany it is a common proverb "that they neede neither Physition to cure their inward diseases nor Chirurgion to helpe them of any wound or sore that have this Bugle and Sanicle at hand by them to use." That this is equally true of the herb self-heale. That country-folk use sanicle to anoint their hands "when they are chapt by the winde." That goat's rue is good for fattening hens. That Herbe True love taken every day for twenty days will help those "that by witchcraft (as it is thought) have become half foolish to become perfectly restored to their former good estate." That the best starch is made from the root of cuckoo-pint, and that in former dayes when the making of our ordinary starch "was not knowen or frequent in use; the finest Dames used the rootes hereof to starch their linnen, which would so sting, exasperate and choppe the skinne of their servants' hands that used it, that they could scarce get them smooth and whole with all the nointing they could doe before they should use it againe." That the root of this same herb, cut small and mixed with a sallet of white endive or lettice, is "an excellent dish to entertain a smell-feast or unbidden unwelcome guest to a man's table, to make sport with him and drive him from his too much boldnesse; or the pouder of the dried roote strawed upon any daintie bit of meate that may be given him to eate; for either way within

PORTRAIT OF JOHN PARKINSON FROM THE TITLE-PAGE OF THE "PARADISUS" (1629)

a while after the taking of it, it will so burne and pricke his mouthe that he shall not be able either to eate a bit more or scarce to speak for paine and so will abide untill there be some new milk or fresh butter given, which by little and little will take away the heate and pricking and restore him againe." That another "good jest for a bold unwelcome guest" is to infuse nightshade in a little wine for six or seven hours and serve it to the guest, who then "shall not be able to eat any meate for that meale nor untill he drinks some vinegar which will presently dispell that qualitie and cause him to fall to his viands with as good a stomach as he had before." That sufferers from toothache should rub the bruised root of crowfoote on to their fingers; by causing "more paine therein than is felt by the toothach it taketh away the pain." That the juice of fumitory, if dropped in the eyes, will take away the redness and other defects, "although it procure some paine for the present and bringeth forth teares." That the hunters and shepherds of Austria commend the roots of the supposed wolf's-bane "against the swimming or turning in the head which is a disease subject to those places rising from the feare and horroure of such steepe downfalls and dangerous places which they doe and must continually passe." That scabious, if bruised and applied "to any place wherein any splinter, broken bone, or any such like thing lyeth in the flesh doth in short time loosen it and causeth it to be easily drawen forth." That butcher's broom was used in olden times to preserve "hanged meate" from being eaten by mice and also for the making of brooms, "but the King's Chamber is by revolution of time turned to the Butcher's stall, for that a bundle of the stalkes tied together serveth them to cleanse their stalls and from thence have we our English name of Butcher's broom." That the down of swallow-wort "doth make a farre softer stuffing for cushions or pillowes or the like than Thistle downe which is much used in some places for the like purpose." That, if ivory is boiled with mandrake root for six hours, the ivory will become so soft "that it will take what form or impression you will give it." That fresh elder flowers, hung in a vessel of new wine and pressed every evening for seven nights together, "giveth to the wine a very good relish and a smell like Muscadine." That the moth mullein is of no use except that it will attract moths wherever it is laid. That if pennyroyal is put into "unwholesome and stinking waters that men must drinke (as at sea in long voyages) it maketh them the less hurtful." And to conclude, it is from Parkinson we learn that "Queen Elizabeth of famous memorie did more desire medowsweet then any other sweete herbe to strewe her chambers withall."

CHAPTER VII

THE LATER SEVENTEENTH-CENTURY HERBALS

"Come into the fields then, and as you come along the streets, cast your eyes upon the weeds as you call them that grow by the walls and under the hedge-sides."—W. COLES, *The Art of Simpling*, 1656.

THE later seventeenth-century herbals are marked by a return to the belief in the influence upon herbs of the heavenly bodies, but it is a travesty rather than a reflection of the ancient astrological lore. The most notable exponent of this debased lore was the infamous Nicholas Culpeper, in whom, nevertheless, the poor people in the East End seem to have had a boundless faith. It is impossible to look at the portrait of that light-hearted rogue without realising that there must have been something extraordinarily attractive about the man who was the last to set up publicly as an astrologer and herb doctor. He was the son of a clergyman who had a living somewhere in Surrey. After a brief time at Cambridge he was apprenticed to an apothecary near St. Helen's, Bishopsgate, and shortly afterwards set up for himself in Red Lion Street, Spitalfields, as an astrologer and herbalist. Culpeper was a staunch Roundhead and fought in at least one battle. All through the war, however, he continued his practice and he acquired a great popularity in the East End of London. In 1649 he issued his *Physical Directory*, which was a translation of the *London Dispensatory*. This drew down on him the fury of the College of Physicians, and the book was virulently attacked in a broadside issued in 1652, entitled "A farm in Spittlefields where all the knick-knacks of astrology are exposed to open sale." By this time his works were enjoying an enormous sale. No fewer than five editions of his *English Physician Enlarged* appeared before 1698, and it was reissued even as late as 1802 and again in 1809. There is a vivid description of Culpeper in *The Gentleman's Magazine* for May 1797 :—

"He was of a middle stature, of a spare lean body, dark complexion, brown hair, rather long visage, piercing quick eyes, very active and nimble. Though of an excellent wit, sharp fancy, admirable conception and of an active understanding, yet occasionally inclined to melancholy, which was such an extraordinary enemy to him that sometimes wanting company he would seem like a dead man. He was very eloquent, a good orator though very conceited and full of jest, which was so inseparable to him that in his most serious writings, he would mingle matters of levity and extremely please himself in so doing. Though his family possessed considerable property it appears he was exceedingly restricted in his pecuniary concerns, which probably was the cause of his early leaving the University, as he observes; though his mother lived till he was twenty-three years of age and left him well provided, yet he was cheated or nearly spent all his fortune in the outset of life. Another author observes it is most true that he was always subject to a consumption of the purse, notwithstanding the many ways he had to assist him. His patrimony was also chiefly consumed at the University. Indeed he had a spirit so far above the vulgar, that he condemned and scorned riches any other way than to make them serviceable to him. He was as free of his purse as of his pen. . . . He acknowledged he had many pretended friends, but he was rather prejudiced than bettered by them, for, when he most stood in need of their friendship and assistance they most of all deceived him."

Culpeper wrote a number of medical works which do not concern us here, but his name will always be associated with his Herbal. His reason for having written it he affirms to be that, of the operation of herbs by the stars he found few authors had written, "and those as full of nonsense and contradiction as an Egg is full of meat. This not being pleasing and less profitable to me, I consulted with my two brothers Dr. Reason and Dr. Experience and took a voyage to visit my Mother Nature, by whose advice together with the help of Dr. Diligence I at last obtained my desire and, being warned by Mr. Honesty (a stranger in our days) to publish it to the world, I have done it." It is impossible to read any part of this absurd book without a vision arising of the old rogue standing at the street corner and not only collecting but holding an interested crowd of the common folk by the sort of arguments which they not only understand but appreciate. In his preface he warns his readers against the false copies of his book " that are printed of that letter the small Bibles are printed with . . . there being twenty or thirty gross errors in every sheet." He is withering in his criticism of those who quote old authors as authorities. " They say Reason makes a man differ from a beast; if that be true, pray what are they that instead of Reason for their judgment quote old authors? " In his preface, as throughout his book, he affirms his belief in the connection between herbs and stars. Diseases, he asserts, vary according to the motions of the stars, " and he that would know the reason for the operation of the herbs must look up as high as the stars. It is essential to find out what planet has caused the disease and then by what planet the afflicted part of the body is governed. In the treatment of the disease the influence of the planet must be opposed by herbs under the influence of another planet, or in some cases by sympathy, that is each planet curing its own disease." Elsewhere he directs that plants must always be picked according to the planet that is in the ascendant. Culpeper asserts that herbs should be dried in the sun,[1] his ingenious reasoning being

[1] John Archer (one of the Physicians in Ordinary to Charles II.) also asserts in his *Compendious Herbal* (1673) that " the Sun doth not draw away the Vertues of Herbs, but adds to them." Archer gives full astrological directions for the gathering of herbs:—

" I have mentioned in the ensuing Treatise of Herbs the Planet that Rules every Herb for this end, that you may the better understand their Nature

this :—" For if the sun draw away the virtues of the herb it must needs do the like by hay, which the experience of every farmer will explode for a notable piece of nonsense." He also pours scorn on those who say that the sap does not rise in the winter. Here his argument is even more remarkable, and yet one cannot help realising how effectual it would be with the class of folk with whom he dealt. " If the sap fall into the roots in the fall of the leaf and lies there all the winter then must the root grow all the winter, but the root grows not at all in the winter as experience teaches, but only in the summer. If you set an apple kernel in the spring you shall find the root to grow to a pretty bigness in the summer and be not a whit bigger next spring. What doth the sap do in the root all the winter that while? Pick straws? 'Tis as rotten as a rotten post." He gives as his own version of what happens to the sap that " when the sun declines from the tropic of cancer, the sap begins to congeal both in root and branch. When he touches the tropic of capricorn and ascends to uswards it begins to wax thin again." One cannot help suspecting that Culpeper knew perfectly well what nonsense he was talking, but that he also realised how remunerative such nonsense was and how much his customers were impressed by it. In his dissertation on wormwood one

---

and may gather them when they are in their full strength, which is when the Planet is especially strong, and then in his own Hour gather your Herb; therefore that you may know what hour belongs to every Planet take notice that Astrologers do assign the seven days of the week to the seven planets, as to the Sun or ☉ Sunday; to the Moon or ☽ Monday; to Mars or ♂ Tuesday; to Mercury or ☿ Wednesday; to Jupiter or ♃ Thursday; to Venus or ♀ Friday; to Saturn or ♄ Saturday. And know that every Planet governs the first Hour after Sun Rise upon his day and the next Planet to him takes the next Hour successively in this order, ♄, ♃, ♂, ☉, ♀, ☿, ☽ ♄ ♃. So be it any day every Seventh Hour comes to each Planet successively, as if the day be Thursday then the first hour after Sun Rising is Jupiter's, the next ♂, the next ☉, next ♀. So on till it come to ♃ again. And if you gather Herbs in their Planetary Hour you may expect to do Wonders, otherwise not; to Astrologers I need say nothing; to others this is as much as can easily be learnt."—*The Compendious Herbal*, by John Archer, One of his Majesties Physicians in Ordinary.

NICHOLAS CULPEPER FROM " THE ENGLISH PHYSICIAN ENLARGED "

feels that he was writing with his tongue in his cheek, especially in the conclusion, which is as follows :—

"He that reads this and understands what he reads hath a jewel of more worth than a diamond. He that understands it not is as little fit to give physick. There lies a key in these words, which will unlock (if it be turned by a wise hand) the cabinet of physic. I have delivered it as plain as I durst . . . thus shall I live when I am dead. And thus I leave it to the world, not caring a farthing whether they like it or dislike it. The grave equals all men and therefore shall equal me with all princes. . . . Then the ill tongue of a prating fellow or one that hath more tongue than wit or more proud than honest shall never trouble me. Wisdom is justified of her children. And so much for wormwood."

Less popular than Culpeper's numerous writings, but far more attractive and altogether of a different stamp, are Coles's two books, *Adam in Eden* and *The Art of Simpling*. The title of the latter runs thus :—

"The Art of Simpling. An Introduction to the Knowledge and Gathering of Plants. Wherein the Definitions, Divisions, Places, Descriptions, Differences, Names, Vertues, Times of flourishing and gathering, Uses, Temperatures, Signatures and Appropriations of Plants are methodically laid down. London. Printed by J. G. for Nath. Brook at the Angell in Cornhill. 1656."

The preface is quaint and so typical of the spirit of the later seventeenth-century herbals that I transcribe a good deal of it :—

"What a rare happiness was it for Matthiolus that famous Simpler, to live in those days wherein (as he himself reports) so many Emperors, Kings, Arch-Dukes, Cardinalls and Bishops did favour his Endeavour, and plentifully reward him ! Whereas in our times the Art of Simpling is so farre from being rewarded, that it is grown contemptible and he is accounted a simple fellow, that pretends to have any skill therein. Truly it is to be lamented that the men of these times which pretend to so much Light should goe the way to put out their owne Eyes, by trampling upon that which should preserve them, to the great discouragement of those that have any mind to bend their Studies this way. Notwithstanding, for the good of my Native Countrey, which everyone is obliged to serve upon all occasions of advantage and in pitty to such Mistakers, I have painfully endeavoured plainly to demonstrate the way of attaining this necessary Art, and the usefulnesse of it, in hopes that this Embryo thrown thus into the wide world, will fall into the Lap of some worthy persons that will cherish it, though I knew not any to whose protection I might commend it. However I have adventured it abroad, and to expresse my reall affection to the publick good have in it communicated such Notions, as I have gathered, either from the reading of Severall Authors, or by conferring sometimes with Scholars, and sometimes with Countrey people; To which I have added some Observations of mine Owne, never before published: Most of which I am confident are true, and if there be any that are not so, yet they are pleasant."

There is something very attractive in the last inconsequent remark!

Coles deals mercilessly with old Culpeper. "Culpeper," he says, "(a man now dead and therefore I shall speak of him as modestly as I can, for were he alive I should be more straight with him), was a man very ignorant in the forme of Simples. Many Books indeed he hath tumbled over, and transcribed as much out of them as he thought would serve his turne (though many times he were therein mistaken) but added very little of his own." He even comments on the fact that either Culpeper or his Printer cannot spell aright—"sure he or the Printer had not learned to spell."

The Doctrine of Signatures he accepts unquestioningly. "Though Sin and Sathan have plunged mankinde into an

Ocean of Infirmities Yet the mercy of God which is over all his Workes Maketh Grasse to grow upon the Mountaines and Herbs for the use of Men and hath not onely stemped upon them (as upon every man) a distinct forme, but also given them particular signatures, whereby a Man may read even in legible Characters the Use of them. Heart Trefoyle is so called not onely because the Leafe is Triangular like the Heart of a Man, but also because each leafe contains the perfect Icon of an Heart and that in its proper colour viz a flesh colour. Hounds tongue hath a forme not much different from its name which will tye the Tongues of Hounds so that they shall not barke at you : if it be laid under the bottomes of ones feet. Wallnuts bear the whole Signature of the Head, the outwardmost green barke answerable to the thick skin whereunto the head is covered, and a salt made of it is singularly good for wounds in that part, as the Kernell is good for the braines, which it resembles being environed with a shell which imitates the Scull, and then it is wrapped up againe in a silken covering somewhat representing the *Pia Mater*."

Of those plants that have no signatures he warns the reader not to conclude hastily that therefore they have no use. "We must cast ourselves," he says, "with great Courage and Industry (as some before us have done) upon attempting the vertues of them, which are yet undiscovered. For man was not brought into the world to live like an idle Loyterer or Truant, but to exercise his minde in those things, which are therefore in some measure obscure and intricate, yet not so much as otherwise they would have been, it being easier to adde than invent at first." He then gives his own curious but naïvely interesting theory of plants "commonly accounted useless and unprofitable." "They would not be without their use," he argues, "if they were good for nothing else but to exercise the Industry of Man to weed them out who, had he nothing to struggle with, the fire of his Spirit would be halfe extinguished in the Flesh." After pointing out that weeding them out is in itself excellent exercise, he proceeds :—"But further why may not poysonous plants draw to them all the maligne juice and nourishment that the other may be more pure and refined, as well as Toads and other poysonous Serpents licke the venome from the Earth? . . . So have I seen some people when they have burned their fingers to goe and burne them again to fetch out the fire. And why may not one poyson fetch out another as well as fire fetch out fire?" "For should all things be known at once," he wisely concludes, "Posterity would have nothing left wherewith to gratifie themselves in their owne discoveries, which is a great encouragement to active and quick Wits, to make them enquire into those things which are hid from the eyes of those which are dull and stupid."

Coles's *Art of Simpling* is the only herbal which devotes a chapter to herbs useful for animals—"Plants as have operation upon the bodies of Bruit Beasts." This chapter is full of curious folk lore. He gives the old beliefs that a toad poisoned by a spider will cure itself with a plantain leaf; that weasels when about to encounter a serpent eat rue; that an ass when it feels melancholy eats asplenium; that wild goats wounded by arrows cure themselves with dittany; that the swallow uses celandine ("I would have this purposely planted for them," he adds); that linnet and goldfinch (and have any birds brighter eyes?) constantly repair their own and their young one's eyesight with eyebright; that if loosestrife is thrown between two oxen when they are fighting they will part presently, and being tied about their necks it will keep them from fighting; that cocks which have been fed on garlick are "most stout to fight and so are Horses"; that the serpent so hates the ash tree "that she will not come nigh the shadow of it, but she delights in Fennel very much, which she eates to cleer her eyesight;" that, if a garden is infested with moles, garlic or leeks will make them "leap out of the ground presently." Perhaps the most remarkable effects of herbs are the two following. "Adders tongue put into the left eare of any Horse will make him fall downe as if he were dead, and when it is taken out againe, he becomes more lively than he was before." And "if Asses chance to feed much upon Hemlock, they will fall so fast asleep that they will seeme to be dead, in so much that some thinking them to be dead indeed have flayed off their skins, yet after the Hemlock had done operating they have stirred and wakened out of their sleep, to the griefe and amazement of the owners."

There is one chapter—"Of plants used in and against Witchcraft"—in which, amongst other things, we learn that the ointment that witches use is made of the fat of children, dug up from their graves, and mixed with the juice of smallage, wolfsbane and cinquefoil and fine wheat flour; that mistletoe, angelica, etc. were regarded as being of such sovereign power against witches that they were worn round the neck as amulets. Also, that in order to prevent witches from entering their houses the common people used to gather elder leaves on the last day of April and affix them to their doors and windows. "I doe not desire any to pin their Faiths upon these reports," says Coles, "but only let them know there are such which they may believe as they please." "However," he concludes, "there is no question but very wonderful effects may be wrought by the Vertues which are enveloped within the compasse of the green mantles wherewith many Plants are adorned."

Coles, nevertheless, treats with scorn, and by arguments peculiarly his own, the old belief in the connection between the stars and herbs. "It [the study of herbs] is a subject as antient as the Creation, yea more antient than the Sunne or the Moon, or Starres, they being created on the fourth day whereas Plants were the third. Thus did God even at first confute the folly of those Astrologers who goe about to maintaine that all vegetables in their growth are enslaved to a necessary and unavoidable dependence on the influences of the starres; whereas Plants were even when Planets were not." In another passage, however, he writes, "Though I admit not of Master Culpeper's Astrologicall way of every Planets Dominion over Plants, yet I conceive that the Sunne and Moon have generall influence upon them, the one for Heat the other for Moisture; wherein the being of Plants consists."

The most attractive parts of the *Art of Simpling* are the chapters devoted to the "Joys of Gardening." Coles tells us that "A house, though otherwise beautifull, if it hath no garden is more like a prison than a house." Of what he has to say about gardens and the happiness to be found in gardening I quote much because it is all so pleasant.

"That there is no place more pleasant [than a garden] may appear from God himselfe, who after he had made Man, planted the Garden of Eden, and put him therein, that he might contemplate the many wonderful Ornaments wherewith Omnipotency had bedecked his Mother Earth. . . . As for recreation, if a man be wearied with over-much study (for study is a weariness to the Flesh as Solomon by experience can tell you) there is no better place in the world to recreate himself than a Garden, there being no sence but may be delighted therein. If his sight be obfuscated and dull, as it may easily be, with continuall poring, there is no better way to relieve it, than to view the pleasant greennesse of Herbes, which is the way that Painters use, when they have almost spent their sight by their most earnest contemplation of brighter objects: neither doe they onely feed the Eyes but comfort the wearied Braine with fragrant smells. The Eares also (which are called the Daughters of Musick, because they delight therein) have their recreation by the pleasant noise of the warbling notes, which the chaunting birds accent forth from amongst the murmuring Leaves. . . ."

"Of the profits" [of a garden] he says, "First for household occasions, for there is not a day passeth over our heads but we have of one thing or other that groweth within their circumference. We cannot make so much as a little good Pottage without Herbes, which give an admirable relish and make them wholsome for our Bodies. . . . Besides this inestimable Profit there is another not much inferior to it, and that is the wholsome exercise a man may use in it. . . . If Gentlemen which have little else to doe, would be ruled by me, I would advise them to spend their spare time in their gardens, either in digging, setting, weeding or the like, then which there is no better way in the world to preserve health. If a man want an Appetite to his Victuals the Smell of the Earth new turned up by digging with a spade will procure it,[1] and if he be inclined to a Consumption it will recover him.

"Gentlewomen if the ground be not too wet may doe themselves much good by kneeling upon a Cushion and weeding. And thus both sexes might divert themselves from Idlenesse and evill company, which oftentimes prove the ruine of many ingenious people. But perhaps they may think it a disparagement to the condition they are in; truly none at all if it were but put in practise. For we see that those fashions which sometimes seem ridiculous if once taken up by the gentry cease to be so." He quotes the Emperor Diocletian, who "left for a season the whole Government of the Empire and forsaking the Court betook himself to a meane House with a Garden adjoyning, wherein with his owne hands, he both sowed set and weeded the Herbes of his Garden which kinde of life so pleased him, that he was hardly intreated to resume the Government of the Empire." "By this time," he concludes, "I hope you will think it no dishonour to follow the steps of our grandsire Adam, who is commonly pictured with a Spade in his hand, to march through the Quarters of your Garden with the like Instrument, and there to rectify all the disorders thereof, to procure as much as in you lyes the recovery of the languishing Art of Simpling, which did it but appeare in lively colours, I am almost perswaded it would so affect you that you would be much taken with it. There is no better way to understand the benefit of it, than by being acquainted with Herballs and Herbarists and by putting this Gentle and ingenious Exercise in practise, that so this

[1] In this connection he quotes Dr. Pinck, Warden of New College, Oxford, who, when he was "almost fourscore yeares old, would rise very betimes in the morning and going into his Garden he would take a Mattock or Spade, digging there an hour or two, which he found very advantageous to his health."

part of knowledge as well as others, may receive that esteem and advancement that is due to it, to the banishment of Barbarisme and Ignorance which begin again to prevaile against it."

The real descendants, so to speak, of the herbal are the quaint old still-room books, many of which survive not only in museums and public libraries, but also in country houses. These still-room books, which are a modest branch of literature in themselves, are more nearly akin to herbals than to cookery books, with which they are popularly associated. For they are full of the old herb lore and of the uses of herbs in homely medicines. It must be remembered that even as late as the sixteenth and seventeenth centuries every woman was supposed to have some knowledge of both the preparation and the medicinal use of herbs and simples. When the herbal proper ceased and the first books on botany began to make their appearance the old herb lore did not fall into disuse, and the popularity of the still-room books in which it was preserved may be gathered from the fact that one of the first of these to be printed—*A Choice Manual of rare and select Secrets in Physick & Chirurgerie Collected and practised by the Countesse of Kent*[1] (*late dec'd*) —went through nineteen editions. There are some old books which merely inspire awe, for one feels that they have always lived in dignified seclusion on library shelves and have been handled only by learned scholars. But there are others whose leaves are so be-thumbed and torn that from constant association with human beings they seem to have become almost human themselves. Of this type are these old still-room books. They were an integral part of daily life and their worn pages bear mute witness to the fact.

One of the most interesting is the Fairfax still-room book.[2] Its first owner was probably Mary Cholmeley, born during the

[1] Published 1651. The earliest copy in the British Museum is the second edition, 1653.

[2] See *Arcana Fairfaxiana.*

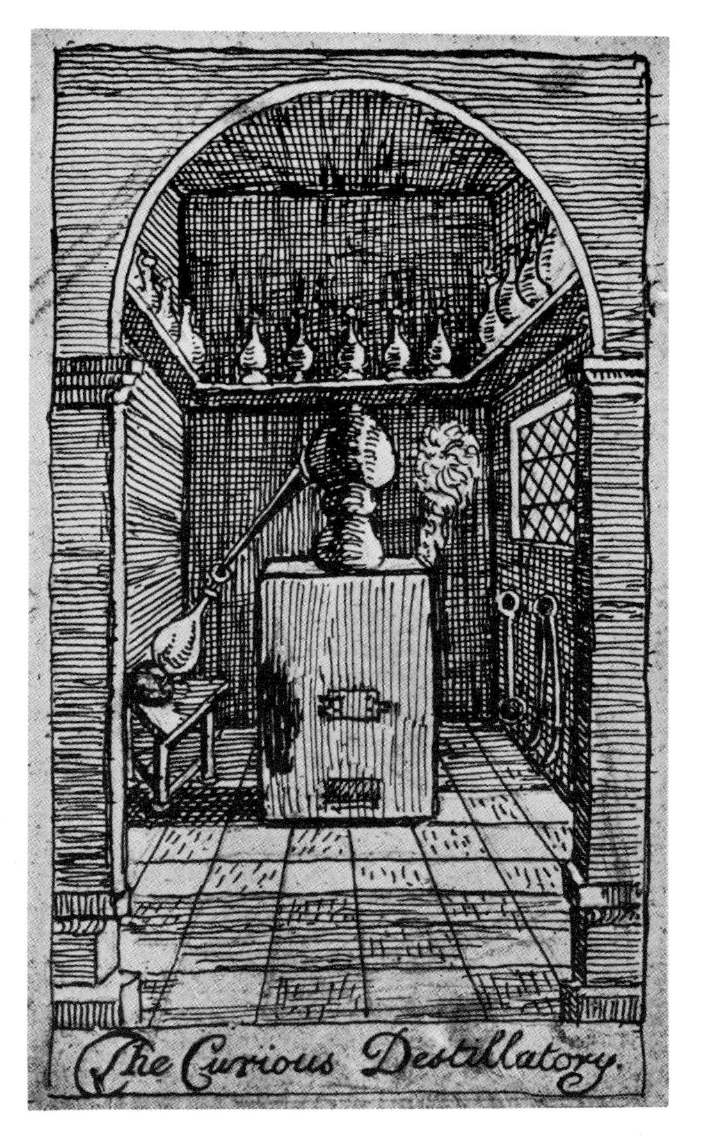

FRONTISPIECE OF "THE CURIOUS DESTILLATORY," BY THOMAS SHIRLEY, M.D., PHYSICIAN IN ORDINARY TO HIS MAJESTY (1677)

closing years of Elizabeth's reign, and married in 1626 to the Rev. the Hon. Henry Fairfax (uncle of the great Parliamentarian General—Lord Fairfax).[1] In common with the majority of MS. still-room books, the Fairfax volume contains much that has no immediate connection with a still-room, but is full of human interest. It is a curious medley of culinary recipes, homely cures, housewifely arts such as bleaching, dyeing, brewing and preserving, to say nothing of hastily scribbled little notes regarding lost linen (including no fewer than "xxiii handkerchares!") and the number of fowls, etc., in the poultry-yard. This last entry, which runs all down one side of a page, is as follows: "I Kapon, XVI Torkies, XVIII dowkes, IIII henes, II cokes, X chekins, X giese, IV sowes."

But the most charming entry of all is: "A note of Mistress Barbara her lessons on ye virginalle which she hath learned and can play them," followed by a list of songs, the majority of which have the entry "Mr. Bird" beside them. William Bird was organist to Queen Elizabeth, and he presumably was "Mistress Barbara's" music-master. She apparently also had lessons from Dr. Bull, then at the height of his fame, for his name appears in connection with some of the items. Amongst the songs we find "My trew Love is to ye grene Wood gon," and there are quite a number of dances—pavanes and courantes—which she played. One feels very sure that "Mistress Barbara" was a fascinating person, but she could not have been more lovable than her sister Mary, who married Henry Fairfax. A love-letter, written in Charles I.'s reign, is doubtless quite out of place in a book on old herbals, but I cannot refrain from quoting the following, written by Mary to her husband about six years after their marriage, because it very clearly reveals the character of one of the many types of women who wrote these still-room books.

[1] Lord Fairfax had only a daughter (who married the Duke of Buckingham), and the son of Henry and Mary Fairfax succeeded to the title.

"MY EVER DEAREST LOVE,

"I received a letter and horse from Long on Thursday (Jan. 31) and will use meine [endeavour] to send Procter's horse to Denton. I did nott so much rejoys att thy safe passage as at that Bleised and al suficiente gide whoss thou art, and whom I know thou truely sarves yt hath for a small time parted us, and I fearmly hope will give us a joyfull meeting. Dear heart, take eassy jernays and preferr thy owne heilth before all other worldly respects whatsoever. . . . I pray y[u] beg a blessing for us all, for I must needs comitt y[u] to his gracious protection yt will never fail us nor forsake us. Thine ever,

"MARY FAIRFAX.

"*Ashton, February* 2, 1632."

I quote only three recipes from this attractive MS.: "A Bath for Melancholy," "Balles for the face" and "For them theyr speech faileth."

"To make a bath for Melancholy. Take Mallowes, pellitory of the wall, of each three handfulls; Camomell flowers, Mellilot flowers, of each one handfull; hollyhocks, two handfulls; Isop one greate handfull, senerick seede one ounce, and boil them in nine gallons of Water untill they come to three, then put in a quart of new milke and go into it bloud warme or somthing warmer."

"Balles for the face. Take greate Allecant reasons [raisins] a quarter of a pounde, stone them but wash them not and beate them in a morter very fine, take as many almonds, not Jordans, but of ye comon sort and blanch them and drye them in a cloth very well and beate them in a stone morter also very fine, when you have done thus to them bothe, mingle them bothe together and beate them againe, and putt to it half a quarter of a pounde of browne leavened bread, wheaten bread, and beate them altogeather and mingle them well togeather and then take it and make it in little balles and then wash yor face at night with one of them in fayre water. Yf you will have this only to wash yor hands put in a little Venice soape but putt none of that in for youre face."

"For them theyr speech faileth. Take a handfull of ye cropps of Rosemary, a handfull of sage and a handfull of Isop and boile them in malmsey till it be soft, then put them into Lynen clothes and laye about the nape of the neck and the pulses of the armes as whott [hot] as it may be suffred daily, as it shal be thought mete and it will help it by God's grace. For the same. Take staves acre and beate it and sowe it in a linnen cloth and make a bagg noe bigger than a beane; if he can chow it in his mouth lett hym, if not then lay it upon his tongue."

To the modern mind the medical recipes to be found in these still-room books sound truly alarming, but in *The Lady Sedley her Receipt book* they are not more so than the prescriptions which were contributed by the most eminent physicians of that day. In his paper [1] on this MS. Dr. Guthrie quotes many of these recipes, amongst them one from the famous Dr. Stephens,[2] so frequently quoted by Sir Kenelm Digby and in other still-room books of the period. In Lady Sedley's book his recipe is introduced thus: "A copy to make the sovreigns't water that ever was devised by man, which Dr Stephens a physician of great cuning and of long experience did use and therewith did cure many great cases, and all was kept in secret until a little before his death; when the Archbishop of Canterbury got it from him." Amongst the other contributors to this MS. were no fewer than three of the doctors who attended Charles II. in his last illness, and if they gave the king even in a mild form medicines resembling those we find in this book, Macaulay's description that "they tortured him for some hours like an Indian at the stake" can hardly have been exaggerated.

[1] Proceedings of the Royal Society of Medicine, 1913.

[2] Dr. Stephens was the author of the *Catalogue of the Oxford Botanical Gardens.*

There is a "Receipt for Convulsion Fitts" from Sir Edward Greaves (the first physician to be created a baronet) consisting of peony roots, dead man's skull, hoofs of asses, white amber and bezoar; and the famous Dr. Sydenham contributed a "Prescription for the head" in which, not content with the seventy-two ingredients of which Venice treacle consisted, he added Wormwood, orange peel, angelica and nutmeg. Another distinguished contributor to this MS. was the ill-fated Duke of Monmouth. A prescription for stone from Judge Ellis consisted of Venice turpentine distilled with various herbs and spices in small ale. It was to be made only in June and taken "three days before the full and three days before the change of the Moone" (incidentally a survival of Saxon moon lore), but the Duke of Monmouth's prescription for the same complaint is quite different and is compounded of ripe haws and fennel roots distilled in white wine and taken with syrup of elder. Lady Sedley, the first owner, and presumaby author of the book, was the wife of Sir Charles Sedley, one of Charles II.'s intimate friends and notorious for his mad pranks. Between her husband and her daughter her life must have been almost unbearable, and it is not surprising that the unfortunate woman ended her days in a mad-house.

Of the MS. still-room books in the British Museum undoubtedly the most interesting is *Mary Doggett: Her Book of Receipts*, 1682.[1] On the first page is affixed a note: "This Mary Doggett was the wife of Doggett the Player who left a legacy of a yearly coat and badge to be rowed for."[2] The MS. is beautifully written and contains an astonishing amount of information on every housewifely art, from washing "parti-coloured stockings" to making perfumes and "Sweete Baggs." Indeed the reading of the headlines alone gives one some idea of the multifarious duties of a mistress of a large

[1] Sloane 27466.

[2] The competition for "Doggett's Coat and badge" amongst Thames Watermen still takes place every August.

house in those days. We find—and I quote only a few—recipes "to make morello cherry cakes," "apricock marmalett," "to preserve Cherrys white," "to candy oranges or lemons or any kind of sucketts," "to preserve almonds," "to preserve damsons," "orange butter," "pippin creame," "to make molds for apricock Plumbs," "apricock wine," "to keep cherrys all the year," "to make cowslip wine," "cakes of clove gilly flowers," "curran wine," "grapes in jelly," "cleer cakes of goosberys," "fine cakes of lemons," "to preserve Rasps whole," "to make Lemon Creame," "lemon Syllibub," "orange biskett," "cheese caks of oranges," "to preserve pippins in slices," "to make plumb biskett," "to pickle Quinces," "to preserve Wallnutts," "to preserve double blew violetts for Salletts," "to candy Double marygold, Roses, or any other flowers," "to make good sorrell wine," "sweet powders for linnen," "to perfume gloves after the Spanish maner," "to souse a pigg," "Almond milk," "to pickle cucumbers," "drinks to cause sleep," "snaile broth," "plasters for bruises," "to make pomades" and "past for the hands." The receipts for "A Pomander," for "Balme water," "to dry roses for sweet powder," and "a perfume for a sweet bagg" are particularly attractive, and I give them below.

"A Pomander. Take a quarter of an ounce of Civitt, a quarter and a half-quarter of an ounce of Ambergreese, not half a quarter of an ounce of ye Spiritt of Roses, 7 ounces of Benjamin, allmost a pound of Damask Rose buds cutt. Lay gumdragon in rose water and with it make up your Pomander, with beads as big as nutmegs and color ym with Lamb [*sic*] black; when you make ym up wash your hands wth oyle of Jasmin to smooth ym, then make ym have a gloss, this quantity will make seaven Braceletes."

"A receipt for Balme. Take 6 or 7 handfulls of balme, cut it a little, put it in an Earthen pott wth a handfull of cowslip flowers, green or dry, half an ounce of Mace, a little bruised pow[d]er in ym, 4 quarts of strong ale, let ym stand a night to infuse: in ye morning put it into your still, poure upon it a quart of brandy. Past up your Still; you may draw about 2 quarts of water. Sweeten it with Sugar to your Tast and tye up too pennyworth of Saffron in a ragg, put it into ye water and let it lye till it be colored. Squeeze it out and bottle it for your use."

"To dry Roses for sweet powder. Take your Roses after they have layen 2 or 3 days on a Table, then put them into a dish and sett ym on a chafering dish of Charcole, keeping them stirred, and as you stir ym strew in some powder of orris, and when you see them pretty dry put them into a gally pot till you use them."

"A perfume for a sweet bagg. Take half a pound of Cypress Roots, a pound of Orris, 3 quarters of a pound of Rhodium, a pound of Coriander Seed, 3 quarter of a pound of Calamus, 3 orange stick wth cloves, 2 ounces of Benjamin, and an ounce of Storax and 4 pecks of Damask Rose leaves, a peck of dryed sweet Marjerum, a pretty stick of Juniper shaved very thin, some lemon pele dryed and a stick of Brasill; let all these be powdered very grosely for ye first year and immediately put into your baggs; the next year pound and work it and it will be very good again."

The "Countesse of Kent's" still-room book, which was one of the first to be published, contains more recipes against the Plague than most, and with one of these we find the instruction that it must be taken three times, "for the first helpeth not." Amongst much that is gruesome there is a pleasant recipe entitled "A comfortable cordial to cheer the heart," which runs thus: "Take one ounce of conserve of gilliflowers, four grains of the best Musk, bruised as fine as flower, then put it into a little tin pot and keep it till you have need to make this cordial following: Viz.: Take the quantity of one Nutmeg out of your tin pot, put to it one spoonful of cinnamon water, and one spoonful of the sirrup of gillifloures, ambergreece, mix all these together and drink them in the morning, fasting three or four hours, this is most comfortable." The *chef d'œuvre* of the collection, at least in the author's opinion, is one introduced with this flourish, but it is too long for me to quote more than the comprehensive title:—" The Countesse of Kent's powder, good against all malignant and pestilent diseases, French Pox, Small Pox, Measles, Plague, Pestilence, Malignant or Scarlet Feavers, good against melancholy, dejection of Spirits, twenty or thirty grains thereof being exhibited in a little warm Sack or Hartshorn Jelly to a Man and half as much or twelve grains to a Childe."

Far more attractive than the volume which bears the " Countesse of Kent's " name are the little-known books by Tryon. They are full of discourses and sermons, introduced at the most unexpected moments. Indeed, there are few subjects on which Tryon does not lecture his readers, from giving servants extra work on Sundays by having " greasy platters and Bloody-Bones more on Sunday than any other Day," to sleeping in feather beds. It is interesting to find that women had already taken to smoking in the seventeenth century, and Tryon admonishes them thus:—" Nor is it become infrequent, for women also to smoak Tobacco. Tobacco being an Herb of Mars and Saturn, it hath its fiery Quality from Mars, and its Poysonous fulsome attractive Nature from Saturn : the common use of it in Pipes is very injurious to all sorts of people but more especially to the Female Sex." Tryon seems to have been somewhat of a Socialist, and he takes great delight in commiserating " Lords, Aldermen, the Rich and the Great," who are driven to " heartily envying those Jolley Swains, who feed only with Bread and Cheese, and trotting up to the knees in Dirt, do yet with lusty limbs, and vigorous stomach, and merry Hearts, and undisturbed Heads, whistle out more sollid joys than the others, with all their Wealth and State can purchase."

The most famous of all still-room books was that written by Sir Kenelm Digby, the friend of Kings and philosophers and himself a man of science, a doctor, an occultist, a privateer and a herbalist. Indeed it would be impossible to catalogue his activities, and he has always been recognised as the type *par excellence* of the gifted amateur. Sir Kenelm was the elder son of the Digby who was one of the leaders in the Gunpowder Plot. Himself a man of European reputation, he numbered among his friends Bacon, Ben Jonson, Galileo, Descartes, Harvey and Cromwell. Queen Marie de' Medici was only one of many women who fell in love with him, but his one love was his wife, one of the most beautiful women of her day—Venetia Anastasia Stanley, immortalised by Van Dyck and Ben Jonson. Sir Kenelm Digby was the intimate friend of Charles I. and Henrietta Maria, and after the Restoration he was a prominent figure at the Court of Charles II. When the Royal Society was inaugurated in 1663, he was one of the Council, and his house in Covent Garden was a centre where all the wits, occultists and men of letters forgathered. Aubrey tells us that after the Restoration he lived "in the last faire house westward in the north portico of Covent Garden where my lord Denzill Hollis lived since. He had a laboratory there." [1] One reads so much of the extravagances and excesses of Restoration days that it is all the pleasanter to remember the people of whom little has been written, the thousands of quiet folk who loved their homes and gardens and took delight in simple pleasures. It is of these people Sir Kenelm Digby's book reminds us, and even the names of his recipes are soothing reading—syllabubs, hydromel, mead, quidannies, tansies, slipp-coat-cheeses, manchets, and so forth. Moreover, there is no savour of the shop in these recipes, the book being full rather of flowers and herbs. It is also very leisurely, and in these days that, too, is soothing. Time we frequently find measured thus:—"Whiles you can say the Miserere Psalm very slowly" or "about an Ave Maria while." It takes us back to a simple old world when great ladies not only looked well to the ways of their households, but attended themselves to the more important domestic matters.

[1] This house is to be seen in Hogarth's "Morning."

Sir Kenelm collected these recipes assiduously from his friends, and each housekeeper's pride in her speciality is very evident. To mention only a few of these, we find :—" Scotch Ale from my Lady Holmeby," " A very pleasant drink of Apples," " Master Webb's Ale and Bragot," " Apples in Gelly," " To make Bisket," " Sir Paul Neal's way of making Cider," " My Lord of St. Alban's Cresme Fouettee," " The Queen's Barley Cream," " To pickle capons my Lady Portland's way," " Pickled Champignons," " A Flomery-Caudle," " My Lord Hollis Hydromel," " Master Corsellises Antwerp Meath," " My own considerations for making of Meathe," " Meathe from the Muscovian Ambassadors Steward," " White Metheglin of my Lady Hungerford's which is exceedingly praised," " My Lord of Denbigh's Almond March-pane," " My Lord Lumley's Pease-pottage," " Pease of the seed, buds of Tulips," " A soothing Quiddany or Gelly of the Cores of Quinces," " Sack with clove gilly-flowers," " My Lord of Carlile's Sack-posset," " To make a whip Syllabub," " Sucket of Mallow-stalks," " The Countess of Newport's Cherry Wine." We may forget the recipes themselves, but the memory of them is associated with the fragrance of gilliflowers, roses, cowslips, elder flowers, violets, thyme, marjoram and the like. I give but these few below, and I wish there were space for more; for not only are they excellent in themselves, but, in common with all those in Sir Kenelm Digby's book, they give more, perhaps, of the atmosphere of the old still-rooms than is to be found in any other collection.

" Sweet meat of Apples. My Lady Barclay makes her fine Apple-gelly with slices of John apples. Sometimes she mingles a few pippins with the Johns to make the gelly. But she liketh best the Johns single and the colour is paler. You first fill the glass with slices round-wise cut, and then the Gelly is poured in to fill up the vacuities. The Gelly must be boiled to a good stiffness. Then when it is ready to take from the fire, you put in some juyce of Lemon, and of Orange too, if you like it, but these must not boil; yet it must stand a while upon the fire stewing in good heat, to have the juyces incorporate and penetrate well. You must also put in some Ambergreece, which doth exceeding well in this sweet-meat."

"Wheaten Flommery. In the West Country they make a kind of Flommery of wheat flower, which they judge to be more harty and pleasant then that of Oat-meal, thus; take half, or a quarter of a bushel of good Bran of the best wheat (which containeth the purest flower of it, though little, and is used to make starch), and in a great wooden bowl or pail, let it soak with cold water upon it three or four days. Then strain out the milky water from it, and boil it up to a gelly or like starch. Which you may season with Sugar and Rose or Orange-flower-water and let it stand till it be cold, and gellied. Then eat it with white or Rhenish-wine, or Cream, or Milk, or Ale."

"A Flomery Caudle. When Flomery is made and cold, you may make a pleasant and wholesome caudle of it by taking some lumps and spoonfuls of it, and boil it with Ale and White wine, then sweeten it to your taste with Sugar. There will remain in the Caudle some lumps of the congealed flomery which are not ungrateful."

"Conserve of Red Roses. Doctor Glisson makes his Conserve of red Roses thus: Boil gently a pound of red Rose-leaves (well picked, and the nails cut off) in about a pint and a half (or a little more, as by discretion you shall judge fit, after having done it once; the Doctor's Apothecary takes two pints) of Spring water; till the water have drawn out all the Tincture of the Roses into itself, and that the leaves be very tender, and look pale like Linnen; which may be in a good half hour, or an hour, keeping the pot covered whiles it boileth. Then pour the tincted Liquor from the pale leaves (strain it out, pressing it gently, so that you may have Liquor enough to dissolve your Sugar) and set it upon the fire by itself to boil, putting into it a pound of pure double refined Sugar in small Powder; which as soon as it is dissolved, put into it a second pound, then a third, lastly a fourth, so that you have four pounds of sugar to every pound of Rose-leaves. (The Apothecary useth to put all the four pounds into the Liquor altogether at once.) Boil these four pounds of Sugar with the tincted Liquor, till it be a high Syrup, very near a candy height (as high as it can be not to flake or candy). Then put the pale Rose-leaves into this high Syrup, as it yet standeth upon the fire, or immediately upon the taking it off the fire. But presently take it from the fire, and stir them exceeding well together, to mix them uniformly; then let them stand till they be cold, then pot them up. If you put your Conserve into pots whiles it is yet thoroughly warm, and leave them uncovered some days, putting them in the hot Sun or stove, there will grow a fine candy on the top, which will preserve the conserve without paper upon it, from moulding, till you break the candied crust to take out some of the conserve.

"The colour both of the Rose-leaves and the Syrup about them, will be exceedingly beautiful and red, and the taste excellent, and the whole very tender and smoothing, and easie to digest in the stomack without clogging it, as doth the ordinary rough conserve made of raw Roses beaten with Sugar, which is very rough in the throat. The worst of it is, that if you put not a Paper to lie always close upon the top of the conserve, it will be apt to grow mouldy there on the top; especially après que le pot est entamé."

Under another "conserve of red roses" we find this note:—
"Doctor Bacon useth to make a pleasant Julip of this Conserve of Roses, by putting a good spoonful of it into a large drinking glass or cup; upon which squeeze the juyce made of a Lemon, and slip in unto it a little of the yellow rinde of the Lemon; work these well together with the back of a spoon, putting water to it by little and little, till you have filled up the glass with Spring water: so drink it. He sometimes passeth it through an Hypocras bag and then it is a beautiful and pleasant Liquor."

These still-room books are as much part of a vanished past as the old herb-gardens, those quiet enclosures full of sunlight and delicious scents, of bees and fairies, which we foolish moderns have allowed to fall into disuse. The herb garden was always the special domain of the housewife, and one likes to think of the many generations of fair women who made these gardens their own, tending them with their own hands, rejoicing in their beauty and peace and interpreting in humble, human fashion something of the wonder and mystery of Nature in the loveliness of a garden enclosed. For surely this was the charm of these silent secluded places, so far removed from turmoil that from them it was possible to look at the world with clear eyes and a mind undisturbed by clamour. And what of the fairies in those gardens? We live in such a hurrying, material age that even in our gardens we seem to have forgotten the fairies, who surely have the first claim on them. Does not every child know that fairies love thyme and foxgloves and the lavish warm scent of the old cabbage rose? Surely the fairies thronged to those old herb-gardens as to a familiar haunt. Can you not see them dancing in the twilight?

The dark elves of Saxon days have well-nigh vanished with the bogs and marshes and the death-like vapours which gave them birth. With the passing of centuries the lesser elves have become tiny of stature and friendly to man, warming themselves by our firesides and disporting themselves in our gardens. Perhaps now they even look to us for protection, lest in this age of materialism they be driven altogether from the face of the earth. As early as the twelfth century we find mention of creatures akin to the brownies, whom we all love; for the serious Gervase of Tilbury tells us of these goblins, less than half an inch high, having faces wrinkled with age, and dressed in patched garments. These little creatures, he assures us, come and work at night in the houses of mankind; but they had not lost their impish ways and elvish tricks, "for at times when Englishmen ride abroad in the darkness of night, an unseen Portunos [Brownie?] will join company with the wayfarer; and after riding awhile by his side will at length seize his reins and lead his horse into the slough wherein he will stick and wallow while the Portunos departs with mocking laughter, thus making sport of man's simplicity." Perhaps they still make sport of our simplicity, but we shall be the losers if they vanish altogether from the earth. If in impish mood they lead the wayfarer into sloughs, do not the sheen-bright elves lighten some of the darkest paths of pain which human beings are forced to tread? Are not these Ariel-like creatures links between the flowers of earth which they haunt and the stars of heaven whence they seem to derive their radiance? The fairies have almost deserted us, but perhaps they will one day come back to our gardens and teach us that there is something true, though beyond what we can know, in the old astrological lore of the close secret communion between stars and flowers. Do not flowers seem to reflect in microscopic form those glorious flowers which deck the firmament of heaven? In many flowers there is something so star-like that almost unconsciously our minds connect them with the luminaries in the great expanse above us, and from this it seems but a short step to the belief that there is between them a secret communion which is past our understanding.

> "This is the enchantment, this the exaltation,
> The all-compensating wonder,
> Giving to common things wild kindred
> With the gold-tesserate floors of Jove;
> Linking such heights and such humilities,
> Hand in hand in ordinal dances,
> That I do think my tread,
> Stirring the blossoms in the meadow-grass
> Flickers the unwithering stars." [1]

Mystics of all ages and of all civilisations have felt this secret bond between what are surely the most beautiful of God's creations—flowers and stars; and its fascination is in no small

[1] Francis Thompson, *An Anthem of Earth.*

part due to the exquisite frailty and short-lived beauty of the flowers of earth and the stupendous majesty of the flowers in the heavens, those myriad worlds in whose existence a thousand years is but as a passing dream.

Goddis grace shall euer endure.

(*Inscription at the end of "The vertuose boke Of Dystyllacyon of the waters of all maner of Herbes."* 1527.)

BIBLIOGRAPHIES

I

MANUSCRIPT HERBALS, TREATISES ON THE VIRTUES OF HERBS, ETC.

MANUSCRIPTS WRITTEN IN LATIN AFTER 1400 ARE NOT INCLUDED IN THIS LIST.

9th (?) century. Liber dialogorum Gregorii cum libro medicinali in duabus partibus quarum altera tractat de virtutibus herbarum et "Herbarium" vulgo dicitur altera de virtutibus lapidum.

Hutton 76. Bodleian.

(This is the translation of Gregory's Dialogues made by Bishop Werefirth of Worcester. The MS. formerly belonged to Worcester Cathedral.)

10th century (Lacnunga). Liber medicinalis de virtutibus herbarum.

Harleian 585. British Museum.

10th century. S. Columbarii Epist. versibus Adonicis scripta. Ad frontem prima paginæ hujus codicis scribuntur manu contemporanea quæ dam de virtutibus herbarum et versiculi nonnuli.

Harleian 3091. British Museum.

10th century. Leech Book of Bald.

Royal 12 D. British Museum.

11th century. Peri Didaxeon. (Saxon translation.)

Harleian 6258. British Museum.

11th century. Herbarium Apuleii Platonici quod accepit ab Ascolapio.

Cott. Vit. C. III. British Museum.

11th century. Incipiunt Capites (capita) libri medicinalis.

Payne 62. Bodleian.

(This is a version of Herbarium Apuleii Platonici.)

11th (?) century. De herba Betonica. Apuleius or Antonius Musa.

CLXXXIX. Corpus Christi College, Oxford.

11th century. Herbarium Apuleius.

Ashmole 1431.

11th (?) century. Herbarium Apuleius.

CLXXX. Corpus Christi College, Oxford.

(This copy once belonged to John Holyngborne.)

11th century. Incipiunt nomina multarum rerum Anglice.

Corpus Christi College, Cambridge.

(In the list occurs Nomina herbarum.)

12th (?) century. Dioscorides de virtutibus herbarum.

Jesus College, Cambridge.

(Formerly at Durham.)

12th century. Exceptìones de libro Henrici de herbis variis.
Digby 13 (VIII). Bodleian.

(The above was compiled, according to Leland and others, by Henry of Huntingdon.)

12th century. Herbarium Apuleius.
Harleian 4986. British Museum.

12th century. Herbarium Apuleius.
Harleian 5294. British Museum.

12th century. De virtutibus herbarum.
Sloane 1975. British Museum.

Late 12th century. Imago Medici Conjurantis Herbas.
Harleian 1585. British Museum.

12th century. De viribus herbarum.
Harleian 4346. British Museum.

(In verse, commonly ascribed to Macrus.)

12th century. Macer de viribus herbarum.
Sloane 84. British Museum.

12th century. De viribus herbarum. Liber Omad.
Digby 13 (VII). Bodleian.

(It is not known who Omad was.)

12th century. Macer de virtutibus herbarum.
Digby 4 (XI). Bodleian.

(The last folio is in thirteenth-century hand.)

12th century. Macer de virtutibus herbarum.
Library of Lincoln Cathedral.

12th century. Herbarium Apuleius.
Library of Eton College.

12th or 13th century. De negociis specierum. Inc. circa instans negocium in simplicibus.
Trinity College, Cambridge.

13th century. Epistola antonii muse ad agrippam de herba betonica.
Gonville and Caius College, Cambridge.

13th century. Antonii Musæ libellus de virtutibus herbæ betonicæ.
Ashmole 1462 (VIII). Oxford.

13th century. [Synopsis libelli Antonii Musæ.]
Ashmole 1462 (II).

13th century. Synopsis Herboralii Apuleii.
Ashmole 1462 (III).

13th century. Herboralium Apuleii Platonis.
Ashmole 1462 (IX).

(The names of the herbs are given in English in rubric by a hand of the fourteenth century.)

13th century. Herbarium apuleii platonici.
Gonville and Caius College, Cambridge.

13th century. Aemilii Macri de viribus Herbarum.
Royal 12 B. British Museum.

13th century. Aemilii Macri Carmen de viribus herbarum.
Ashmole (1398. II. v).

13th century. Liber Macri de viribus Herbarum.
Ee. VI. 39. II. Cambridge University Library.

Late 13th century. Aemilii Macri de Herbarum viribus.
(Formerly belonged to St. Augustine's Abbey, Canterbury.)
Royal 12 E. XXIII. (III). British Museum.

Late 13th century. Liber Macri de Naturis herbarum.
Kk. IV. 25. (XVI). Cambridge University Library.

13th (?) century. Liber Macri de viribus herbarum.
438 (III). Corpus Christi College, Cambridge.

13th century. De simplicibus medicinis.
505 (II. 2). Corpus Christi College, Cambridge.

13th century. Poema de virtutibus herbarum macro vulgo adscriptum.
Arundel 283 (I). British Museum.

13th century. Aemilius Macri (Fragment).
Library of Lord Clifden, Lanhydrock, Cornwall.

13th century. Le livre de toutes herbes appele " Circa instans."
Sloane 1977. British Museum.

13th century. Le livre de toutes herbes appele " Circa instans."
Sloane 3525. British Museum.

Incipit [liber de simplicibus medicinis ordine alphabetico qui appellatur] " Circa Instans " Plateārii.
Ashmole 1428 (II).

Late 13th century. Circa Instans.
All Souls College, Oxford.

13th century. De medicinis simplicibus sive de virtutibus Herbarum libellus.
Balliol College, Oxford.

Late 13th century. De simplicibus medicinis.
Corpus Christi College, Cambridge.

13th–14th century. Liber cogitanti michi de virtute simplicium medianarum.
Trinity College, Cambridge.

13th–14th century. Herbarium.
Addit. 22636 (XIII). British Museum.

13th–15th century. [Lines on the virtues of the scabious plant.]
Addit. 33381 (XXXVIII). British Museum.

Late 13th century. De collectione herbarum.
Arundel 369 (II). British Museum.

13th century. De Naturis Herbarum.
Royal 8c IX. (X). British Museum.
(From St. Mary's, Reading.)

13th century. De virtutibus herbarum rhythmice.
Sloane 146 (III). British Museum.

13th century. Præfatiuncula in totum præsens volumen.
Inc. In hoc continentur libri quattuor medicine Ypocrates Platonis Apoliensis urbis de diversis herbis.
Ashmole 1462 (I).

Late 13th century. Hic sunt virtutes scabiose distincte.
Digby 86 (LXXXV). Bodleian.

13th (?) century. De proprietate herbarum.
Laud Latin 86 (XIII). Bodleian.

13th–14th century. Liber qui vocatur " Circa Instans."
Peterhouse College, Cambridge.

13th–14th century. Circa instans Platearii.
Trinity College, Cambridge.

13th–14th century. Versus de Ysope (Hyssop).
Harleian 524 (CLXI). British Museum.

Circ. 1300. De herba Basilisca seu Gentiana.
Harleian 2851 (XXIX). British Museum.
(Written in England.)

13th century. Beginning of a history of trees and plants which ends abruptly on page 3.
Harleian 4751. British Museum.

Late 13th–14th century. [Verses—including 19 lines on various herbs.]
Royal 12 c. VI. (VI). British Museum.
(Belonged to Bury St. Edmunds Abbey.)

Circ. 1360–70. Le liure de herberie en français qui est apele " Circa Instans," translated from Johannes Platearius, *De simplici medicina.*
Bodley 76 I. Bodleian.

14th century. MS on the virtues of herbs.
Library of Eton College.

14th century. Macri pœma de Viribus herbarum; præmittitur tabula.
Harleian 2558 (XXIV). British Museum.

14th century. De viribus herbarum pœma.
Sloane 420 (XL). British Museum.

14th century. De viribus herbarum.
Rawl. C. 630. British Museum.

14th century. Macer de virtutibus herbarum.
Sloane 340 (XV). British Museum.

14th century. De virtutibus herbarum.
Digby 95. Bodleian.

Early 15th (?) century. Macer. Of virtues of herbis.
Hutton 29. Bodleian.

14th century. Aemilii Macri de Herbarum viribus.
Royal 12 B. III (I). British Museum.

14th century. Aemilii Macri Carmen de viribus medicinalibus herbarum cum nominibus earum Anglia explicatis.
Ashmole 1397 (E. XV). Oxford.

14th century. Macer de viribus herbarum.
Ff. VI. 53 (X). Cambridge University Library.

14th century. Macer de Herbarum viribus.
36 (I. i). Emmanuel College, Cambridge.

Early 14th century. Aemilius Macer. Carmen de viribus herbarum.
Arundel 225 (II). British Museum.

14th century. De viribus herbarum.
Harleian 3353 (I). British Museum.

14th century. [De virtutibus Ros marine in English.]
759 (XI). Trinity College, Cambridge.

14th century. Herbal in alphabetical order with descriptions.
Arch. Selden 335. Bodleian.

14th century. On simples. Latin and English.
1398 (III). Trinity College, Cambridge.

Late 14th century. Herbarium.
C. XIII (IV). St. John's College, Oxford.

14th century. Here begynnyt a tretys of diverse herbis and furst of Bytayne (Old English poem of 43 couplets).

Begins—

" To tellyn of bytayne I have grete mynde
And sythen of othur herbys os I fynde.
Furst at bytayne I wyl begynne
Yat many vertues berys wt inne."

Last line—

" Yche stounde whyle it mai on erthe be founde."

Ashmole 1397 (II–IV).

Early 14th century. Experimenta Alberti Magni de herbis lapidibus et animalibus.
Addit. 32622. British Museum (III).

14th century. Secreta fratris Alberti de Colonia ordinis fratrum predicatorum super naturis quarundum herbarum et lapidum et animalium in diversis libris philosophorum reperta et in unum collecta.
Digby 147 (XXIV). Bodleian.

14th century. Secreta fratris Alberti ordinis fratrum predicatorum (i) de herbis xvi (ii) de lapidibus (iii) de animalibus (xviii).
Digby 153 (IX). Bodleian.

14th century. Bartholomæus Anglicus de proprietatibus rerum.
Royal 12 E. III. British Museum.

14th century. Bartholomæi Mini de Senis Tractatus de herbis figuris quam plurimis coloratiō instructus.
Egerton 747 I. British Museum.

14th century. ? Gardener. Of the virtues of the herb rosemary, etc.
In the Earl of Ashburnham's library at Ashburnham Place. 122 (2. II).

14th century. Diversitates herbarum omnium que ad medicinas pertinent.
Addit. 29301 (III). British Museum.

(The above has fine pen-and-ink drawings of 68 English wild plants, with their names written in English. The MS. belonged to the Countess of Hainault, Philippa, Queen of England, and, lastly, to Mr. Pettiford.)

14th century. Herbal.
Dd. VI. 29. VII. University Library, Cambridge.

14th (?) century. List of herbs: English names also given.
198 (III). Gonville and Caius College, Cambridge.

(The above once belonged to John Argenteux, Provost of King's.)

14th (?) century. A list of remedies with English equivalents and marginal additions in another hand.
200. Gonville and Caius College, Cambridge.

14th century. [Recipes in Physicke] Glossary containing many herbs.
Pepys Library 1661. Magdalene College, Cambridge.

14*th century*. Here begynneth medecines gode for divers euelys on mennes bodys be callen erchebysschopes auicenna and ypocras Icoupoñ (? cophon) *i. e.* de and on hole materie aȝen brouȝt and ferst of herbis.

Pepys Library 1661. Magdalene College, Cambridge.

(Various simples are described. After the "vertues of rose maryne" a series of sections in verse written as prose beginning "I wil ȝou tellyn by & bi as I fond wretyn in a book. Þat in borwyng I be took of a gret ladyes prest þat of gret name þe mest." The following sections are on centaurea, solsequium, celidonia, pipernella, materfemia, mortagon, pervinca, rosa, lilium, egrimonye. Ends "Oyle of mustard seed is good for ache and for litarge and it is mad on þe same maner.")

Circ. 1400. A treatise in rhyme on the virtues of herbs.

Sloane 147 (V). British Museum.

It begins—

"Of erbs xxiiij I woll you tell by and by
Als I fond wryten in a boke at I in boroyng toke
Of a gret ladys preste of gret name she barest
At Betony I wol begyn at many vertuos het within."

14*th century*. De virtutibus herbarum quarundam.

Ashmole 1397.

(On the medical uses of some herbs. Begins, "Bytayne and wormewode is gode for woundes.")

14*th* (?) *century*. List of names of herbs in Latin and English.

1377 II. Trinity College, Cambridge.

Begins, "Apium Commune Smalache."

1352. De preperacione herbarum. A treatise on the medicinal qualities of and modes of preparing herbs, quoting Serapion. A short list giving first the Latin and then the Irish name, etc.

23 F. 19. Royal Irish Academy.

14*th century*. Vocabulary of herbs in Latin and Welsh.

Addit. 14912. British Museum.

14*th century*. Meddygon myddfai or the Practice of Physic of the Myddvai Doctors: a collection of Recipes for various diseases and injuries, prognostics, charms, virtues of herbs, etc., by the physicians of Myddvai co. Caermarthen.

Addit. 14912 (I). British Museum.

(In Welsh.)

14*th century*. Nomina herbarum. Latin and English.

Addit. 17866. British Museum.

14*th century*. De virtutibus herbæ.

Arundel 507. British Museum.

(The above once belonged to Richard Seybrok, a monk of Durham.

14*th*–15*th century*. Nomina quarundam . . . plantarum arborum.

Harleian 210 (XI). British Museum.

(In French and English.)

14*th*–15*th century*. Names of herbs in Latin and English.

Harleian 2558 (I). British Museum.

14th century. Herbal. Latin and English.
(Directions in gathering herbs, flowers, roots, etc.)
Sloane 2584. British Museum.

14th century. Liber cinomorum (synonomorum) de nominibus herbarum. (Latin, French, English.)
Bodleian 761.

1360–70. Nomina herbarum. (Latin, French, English.)
Bodleian 761 (VI. B.).

Two texts from this MS. were published by E. Mannele Thompson, *Chronicon Galpedi de Baker de Swynebroke.* Clarendon Press, 1889. He gives a list of the contents of this volume, calling this item fol. 158, " Medicinal notes from Roger Bacon in Latin." Interpolated by fifteenth-century writer in spaces left vacant by the fourteenth-century scribe are many recipes and much astrology.

14th century. Virtues of rosemary in prose and verse.
Digby 95 (VII). Bodleian.

14th century. Of the virtues of herbs.
Digby 95 (VIII). Bodleian.

Late 14th century. Herbarium Anglo-Latinum, with many recipes interpolated in a later hand.
MS. Grearerd. Bodleian.

Late 14th century. Names of herbs in alphabetical order with a few English interpolations. The MS. comes from Llanthony Priory and was given by R. Marchall.
312 (X). Library of Lambeth Palace.

14th century. De simplici medicina John Platearius.
(This MS. is supposed to have belonged to the Countess of Hainault and subsequently to Queen Philippa of Hainault.)
Addit. 29301 (IV). British Museum.

14th century. Nomina Herbarum Medicinalium, with some English and French names.
Phillipps MS. 4047 (II) now in the library of T. Fitzroy Fenwick, Esq., Thirlestaine House, Cheltenham.

14th century. Here ben the virtues of Rosemarye (purporting to be taken from " the litel boke that the scole of Sallerne wroat to the Cuntasse of Henowd and sche sente the copie to hir douȝter Philip the quene of England ").
Inc. " Rosemarye is boþe tre and herbe hoot and drie."
Exp. " Wasche him þerwiþ and he schal be hool."
Royal 17 A. III. (III). British Museum.

1373. Translation of Macer *De viribus herbarum* by John Lelamour, Schoolmaster of Hereford.
Sloane 5. British Museum.

14th century. Particulars of simples arranged under the various months.
754. Trinity College, Cambridge.

14th century. A herbal in Latin and English beginning with Allium.
(Given by Thomas Gale Dean of York.)
759 (VII). Trinity College, Cambridge.

15th century. Aemili Macri de virtutibus herbarum. The names of the plants are explained in English in the margins, and there are also some remedies in English. Ashmole 1481 (III).

15*th century*. Macer. De Virtutibus Herbarum. The English names of the herbs are also given. (Written by Nicholas Kyrkeby of Saint Albans.)
VI. 15. Bishop Cosin's Library, Durham University.

15*th century*. Herbal in three books.
Inc. "Mogworte or brotheworte ys clepid archemisia . . . and this medicine ys a nobil medycyne."
Ends, "Here endeth the third part of Macer. And here begynneth a fewe herbes which Macer foryete noȝt nor thei ben nort founden in his book."
Addit. 37786 (II). British Museum.

15*th century*. The treatise of Macer intitled "De viribus Herbarum," translated into English.
"Here followeth the cunnynge and sage clerk Macer tretynge and opynly shewyth the vertuys worthy and Commendable propyrtes of many & dyuerse herbys and her vertuys of the whyche the firste is mugworte or modirworte."
Sloane 393. British Museum.

15*th century*. The vertuys of Erbys aftyr Galyon Ypocras and Socrates.
Lansdowne 680 I. British Museum.

15*th century*. Here folwythe the vertu of Erbis. Isop is hoot and drie in ij degreis so seith Ipocrace if a man drynke it fastynge.
Ashmole 1477 (III–IV).

15*th century*. Aemilius Macer. Of the virtues of herbs. English translation.
Sloane 140. British Museum.

15*th century*. Aemilius Macer. Of the virtues of herbs. English translation.
Sloane 2269. British Museum.

15*th century*. Aemilius Macer. De virtutibus Herbarum. English translation.
In the library of the Right Hon. Lord Amherst of Hackney at Didlington Hall, Norfolk.

15*th century*. List of herbs in Latin and English.
Sloane 3548. British Museum.

15*th century*. Herbal.

Inc. "Of herbys now I
Will you telle by and by.
As I fynde wryten in a boke
That in borrowyng I betoke
Of a gret ladyes preste," etc.

Expl. "It dryveth away all foul moysteris
And distroyeth venym and wykyd humours
It distroyeth the morfew
And dispoyling to the leper."

Dd. X. 44 (VIII). Cambridge University Library.

15*th century*. An Herbary þe whiche ys draw out of Circa Instans and hyt towcherþ schortlyche þe principal vertuys and þe special effectes of herbis and droggis þt be þe most comyne in use, and her dyvers grees of qualites or yher complexions and her propur and most special kynd of worcheyng.
(At the end of every alphabetical division of this work is left a page or

more, blank, for the purpose of inserting additional matter. There are several additions by old hands. Some additions on the margins have been torn off.)

Ashmole 1443 (IV).

15*th century*. Treatise on herbs. 169 chapters, with table of Contents prefixed.
Inc. " Agnus castus is a herbe that men clep Tutsayne or Park levis."
Arundel 272 (II). British Museum.

15*th century*. An Herbal. Arranged alphabetically to the letter P.
Inc. " Agnus castus is an herbe," etc. Breaks off in " pulegium rurale." (Other copies—both ending with S—are in Addit. 4698, f. 16*b*, and Arundel 272, f. 36.)
Royal 18 A. VI. (VI). British Museum.

15*th century*. A treatise on the virtues of Herbs; beginning " Agnus castus ys Anglice herbe that men cally the tutsayne or ells parkelenus."
Ashmole 1432 (V. i).

Mid 15*th century*. Herbal with book of recipes.
Inc. " Agnus castus is an herbe."
Bodleian 463 (A).

15*th century*. Liber de Herbarum virtutibus.
Inc. " Agnus castus ys an herbe that cleepeth Toussane."
Laud Misc. 553 (i). Bodleian.

15*th century*. An Herbal with the properties of the different herbs in alphabetical order, with a table prefixed.
Inc. " Agnus castus ys an herbe that me clapys Tustans or Porke levys."
329. Balliol College, Oxford.

15*th* (?) *century*. " An English Herbal."
Begins, " Agnus Castus," etc.
Harleian 3840 (II). British Museum.

15*th century*. A treatise on the virtues of herbs.
Begins, " A bed ymade of Agnus Castus."
Sloane 297 (XVIII). British Museum.

15*th century*. Latin-English dictionary of herbs.
Inc. " Alleluya Wodsoure stubwort."
Expl. " Quinquefolium fyveleved gras."
Dd. XI. 45 (XII). Cambridge University Library.

15*th century*. A book of the medical virtues of herbs, described in alphabetical order.
Inc. " Anet ys an herbe that ys clepyt anet oþer dylle."
Expl. " doyth a way the fowȝe or the fragelys."
Ashmole 1447 (IV. i).

15*th century*. " Yes ben ye vertuse of betayn."
Ashmole 1438 (II. vii).

15*th century*. A treatise of the virtues of certain herbs. Begins, " Betaigne is hot and drie in þre degrees, and so seyth Ypocras, and it is an herbe of many faire vertues."
Ashmole 1438 (XXV).

15*th century*. Aemilius Macer. De virtutibus herbarum. (In French, Latin and English.)
Digby 29 (XXXVII). Bodleian.

15th century. Of the virtues of herbs—seemingly out of Macer. The following verse is prefixed:

> "This booke ys drawe be fesyke
> That Macer made for hem that ben seeke
> The vertu of herbis hēt descrieth ryght wel
> And help of mannys helthe every del."

Sloane 963 (XVIII). British Museum.

15th century. Macer on the virtues of herbs.
Inc. "Mugworte or brotheworte is clepid Arthemisia."
Exp. "drynkys juse of thys erbe."
Ee. I. 15 (III*a*). University Library, Cambridge.

15th century. Macer. "Vertues worthe & commendable propertees of many & diverse herbes." In three books.
Rawl. C. 81 (V). Bodleian.

15th century. Part of the poem De virtutibus Herbarum. The English names of plants are occasionally given in the margin. In the volume containing Froucestre's History of the Monastery.
Library of Gloucester Cathedral.

15th century. A treatise of the medical properties of herbs and other simples; arranged alphabetically, being a translation from the treatise of Johannes Platearius, *De medicinis simplicibus.*
Sloane 706 (IV). British Museum.

15th century. English Herbal, Secundum magistrum Gilbertum Kemor, arranged alphabetically.
Sloane 770. British Museum.

15th century. Of the virtues of Rosmaryne.
Inc. "Rosmaryne is both tre and erbe."
Sloane 7 (VI). British Museum.

15th century. The virtues of Rosmaryn.
Inc. "Rosmaryn is bothe tre and herbe."
Sloan 962 (VI). British Museum.

15th century. These ben sum of þe vertues of Rosemary, as the Clerke of Sallerne seyde and wrote tho the Cowntes of Hynde, and sche sende hem tho here dowȝtur Phylype þ^t was weddyde tho þe Kyng of Engelond.
Inc. "Rosmary ys bothe tre and herbe."
Ashmole 1438 (II–XX).

15th (?) century. This is ye lityl boke of ye vertuys of rosmaryn yt ye scole of Salerne gaderyd & compiled at instance of ye Cowntese of Henowde. . . . I danyel bain translatyd into vulgar ynglysch worde for werde as fonde in latyn. (The translator adds that before 1432 Rosemary was unknown in England and that it was first sent from the Countess of Hainault to her daughter Queen Philippa.)
1037 (1) (XIV). Trinity College, Cambridge.

15th or early 16th century. The medical virtues of Rosemary in prose. Begins, "Rosus marinus is called rose mary, the virtue of this herbe is goode." Ends, "ne brennyng of unkynd hete be at þi stomake ne at þ^e hert." (At the foot of page 3 is written "Robert Hychys is the ower of thys boke.")
Ashmole 1379 (I).

15th century. Here is vertues and seltyng of Rosmary by the ij doctours of fysyk followyng. per Galyen and Platery, and a poem beginning "As in a booke wretyne y fownd Of wise doctours in dyvers lond."

Ashmole 1379 (II).

15th century. Here follwyth ye wertues off ye rosses mare.

Inc. "Take rosmare and bynd hem ynne a lynnene clothe."

Exp. "Allsso make a bathe off ye floure and y^t^ wyll make ye yonglyche."

Ashmole 1432 (V. iii).

15th century. The vertu of rose mary. Tak þe flower of þe rose mary and bynd hem.

(The above is part of a series of herbal notes, etc., interspersed by a later hand in the course of and following on a fifteenth-century book of medicine.)

Ashmole 1391 (VIII).

15th century. "Here men may see þe vertus of dyuerse herbes, whiche ben hoot and whiche ben coold, and to how many þinges they arne goode." (Other copies are in Sloane 393, f. 13; 1592, f. 39*b*; 3466, f. 78; Addit. 12056, f. 3; Lansdowne MS. 680, f. 2 and 17 B., XLVIII, f. 2, where, however, the arrangement is somewhat different. On page 2 there is the entry, "This is John Rice is boke, the which cost him xxv d.")

15th century. "Here men may se the vertu of dyverse herbes, and what thei be, and whiche ben hoote and which ben colde. And for howgh many thynges they ben goode."

(This MS. ends abruptly in "Calamynte.")

Ashmole 1444 (I. iii).

15th (?) *century*. "The virtues of diuerse herbes which ben hoote and which ben coolde." (With a large table of Contents prefixed.)

Sloane 393 (I). British Museum.

15th century. Treatise on the virtues of herbs. Begins, "Aristologia rotunda. The virtue of this herbe os Ypocras says."

Sloane 962 (XII). British Museum.

15th century. An Herbary or alphabetical Materia Medica of herbs & other drugs; beginning with Aloen, Aloes, Aurum, and ending with Zelboarium.

Inc. "Aloen. To purge fleume and malancoly and colore."

Exp. "Zelboarium. To moysten and to norschen and to clensen and wyth cold þinges to akelen. Amen."

Ashmole 1481 (II. ii).

15th century. An alphabeticall catalogue of Herbes.

Inc. "Aloen hath virtue to purge flewne."

Ee. I. 13 (I). Cambridge University Library.

15th century. A collection of remedies in English (with additions in other handwritings). Begins with "Aloe" and ends with "verveyn."

609 (II). Gonville and Caius College, Cambridge.

15th century. In Latin and English. Herbal. Aloe—Zucarium, with notes on Egrimonia, Acacia, in Latin, and on Cassia lignea and Castorium in English.

43. Jesus College, Cambridge.

15th century. The makynge of oyles of divers herbys.

905 (II. 4). Trinity College, Cambridge.

15*th century*. These ben the precious watris & vertuous for diverse ejvellys.

Inc. " Water of wormode is gode . . . grete lordes among the Saracens usen to drink hitt."

Addit. 37786 (I). British Museum.

15*th century*. Of the Herb Moon-wort.

Inc. " I schal you tel of an Erbe þat men cal Lunarie,
He ys clepit Asterion; wych ys an Erbe þat men calleth Lunarie."

Harleian 2407 (IX). British Museum.

15*th century*. Virtues of the onion, garlic and pennyroyal.

Begins, " Here beeth þe vertues of the Oynoun."

Royal 17 B. XLVIII. (II). British Museum.

15*th century*. Miscellaneous recipes and extracts from herbals.

Begins, " Rosa rebia [*sic*] ys an herbe that men clepyth rede rosys."

Royal 18 A. VI. (VII). British Museum.

15*th* (?) *century*. A treatise of herbs and the several medicaments compounded from them.

Begins, " The roose as saith the philosopher Plinius hath doble verteus."

Sloane 67 (II). British Museum.

15*th century*. A treatise of herbs, alphabetically arranged. (Imperfect.)

Begins, " Carabana id est wylde hempe."

Sloane 297 (I). British Museum.

15*th century*. A treatise of the temperature and virtues of simples alphabetically arranged.

Sloane 965 (VII). British Museum.

15*th century*. " Here men may se the vertues of herbes."

Bodley 463 (B iii).

15*th century*. Liber de herbarum virtutibus.

Inc. " Here may men se the vertu of herbes which ben hot and which ben colde."

Laud Misc. 553 (II). Bodleian.

15*th century*. Vertues of Herbes.

Inc. Apium is an herbe that men call smallache or marche.

Addit. A. 106 (A. IV). Bodleian.

15*th century*. " Here begynnythe to mak waters of erbys sondry and þer vertues and howe þei schalle be made in stillatorie."

Inc. " In þe fyrst of dyl. The water is of gret vertue."

Ashmole 141 B (II. v).

15*th century*. Instructions for the proper time of gathering simples by name.

Inc. " Medysines ben done, some by leves [som] bi sedis, som by flowres and some bi fretes."

Ashmole 1481 (II. iii). Oxford.

15*th century*. The medical use " Of waters distilled from Sundry plants & flowers."

(The above belonged to Richard Saunders, the Astrologer.)

Ashmole 1489 (II. ii).

15*th century*. Alphabetical Herbary.

Inc. " Agrymonia is an herbe."

Bodley 463 (B. ii).

Late 15th century. Virtues of herbs.
Inc. " Here a man maye see."
Selden, *supra* 75 (E. VI). Bodleian.

Late 15th century. A treatise on the properties of plants, fruits, meat and drinks as food and medicine. (In Welsh.)
Jesus College, Oxford.

15th century. Names of herbs.
(Given by Humphrey Moseley, 1649).
69. Emmanuel College, Cambridge.

15th century. Verses in English and Latin on herbs and spices.
(Given by W. Moore.)
176 (I. 2). Gonville and Caius College, Cambridge.

15th century. Recipes in English and Latin.
(Given by W. Moore.)
230 (II). Gonville and Caius College, Cambridge.

15th century. Herbes for a saled.
(This once belonged to Nicholas Butler.)
414 (*d*). Gonville and Caius College, Cambridge.

15th century. Collection of recipes in English, probably all by John Ardern of Newark. Illustrated with rough coloured drawings of herbs, instruments and patients. It begins, " This is a mirrour of bloodletynge in þe weche þey þt wolen beholden it diligently," etc. There is a recipe in French for Greek fire. *Exp.* " tabula libri Sirurgice." Mag. Joh. Arderne de Newerk.
(Given by Humphrey Moseley, 1649.)
69. Emmanuel College, Cambridge.

15th century. Here begynnythe an herball of namys & vertues of diverse herbys aftyr letterys of the a, b, c, etc.
905 (I). Trinity College, Cambridge.

15th century. Virtues of various plants.
905 (II. 4). Trinity College, Cambridge.

15th century. On the virtues of herbs.
Inc. " This booke is drawe be Fesyk. That Macer made for hem þat been seck. Y^e vertu of herbis it discryeth ryght wel."
1637 (I. i). Trinity College, Cambridge.

1485. A collection of the Latin and English names of plants with their descriptions and medical virtues.
National Library of Wales, Aberystwyth.

15th century. Alphabetical list of herbs. (Names partly in Latin and partly in Irish.)
2306. Royal Irish Academy.

15th century. Alphabetical treatise on herbs and their uses. In Latin and Irish.
1315. Trinity College, Dublin.

15th–16th century. List of plants used in medicine. (In Latin and Irish.)
1334 (V). Trinity College, Dublin.

15th century. Vertues of rose maryne þat er contened & compyled in þis space & ar gadirde out of bukes of gude philosofirs & of oþer wyse clerkes.
V. IV. 1. Durham University, Bishop Cosins Library.

Late 15*th century.* Herbal in Welsh.
In Mr. Wynne's library at Peniarth, Merioneth.

15*th century.* The vertu of Rose-marry & other Secrets.
Harleian 1735 (XII). British Museum.

15*th century.* Verses on the virtues of Rosmaryne.
Sloane 3215. British Museum.

15*th century.* Vertues of the herb betayne.
Rawl. C. 211 (II). Bodleian.

15*th century.* Treatise on the vertues of herbs.
Addit. 12056. British Museum.

15*th century.* Treatise on the vertues of herbs & metals in alphabetical order. In Irish.
Addit. 15403. British Museum.

Late 15*th century.* Herbal.
Inc. Agnus Castus is an herbe.
Harleian 3840 (III). British Museum.

15*th century.* A fragment of a treatise on the virtues of herbs.
Sloane 7 (III). British Museum.

15*th century.* An alphabetical herbal.
Sloane 297 (VII). British Museum.

15*th century.* " Of the vyrtues of the Asche tree," etc.
Sloane 297 (XVII). British Museum.

15*th century.* The first part of an intended complete body of Pharmacy in seven parts. The first part treats of herbs, which are alphabetically arranged in 150 chapters.
Sloane 404 (I). British Museum.

15*th century.* On the virtues of herbs, with recipes for various disorders. The last is a charm " for alle maner woundys."
Sloane 540 (I). British Museum.

15*th century.* For to knowe the ix Sauge levys.
Sloane 706 (VIII). British Museum.

15*th century.* Treatise on the virtues of herbs alphabetically arranged.
Sloane 1088 (I). British Museum.

15*th century.* Herbes necessarie for a Gardyn.
Sloane 120 (I). British Museum.

15*th century.* On the virtues of herbs.
Sloane 2403. British Museum.

15*th century.* Poem on the virtues of herbs.
Sloane 2457. British Museum.

15*th century.* Treatise on the virtues of herbs.
Sloane 2460. British Museum.

Early 15*th century.* A fewe othre dyverse herbes with her vertues wich be not yfound in the bokes of Macer.
Rawl. C. 212 (II). Bodleian.

15*th* (?) *century.* A treatise on medicinal herbs. (In Irish.)
Royal Irish Academy, 23 H 19.

15*th century.* A fragment of a treatise on the medicinal properties of herbs. (In Irish.)
Royal Irish Academy, 2306.

15th (?) *century.* A treatise on herbs and their medicinal qualities and the mode of preparing and administering them. (In Irish.)
Royal Irish Academy, 2395.

15th–16th century. Alphabetical list of plants used in medicine and the manner of preparing them. (In Latin and Irish.)
1334 (II). Trinity College, Dublin.

1415. Alphabetical list of plants used in medicine. At the end is the transcriber's name, "Aedh Buide O'Leigin," and the date 1415. Also the name of the person from whom the original MS. was purchased—"Tad hg O'Cuinn bachelor in physic." (In Irish.)
1343 (II). Trinity College, Dublin.

15th century. A dictionary of herbs in Latin and English.
In the Marquis of Bath's library at Longleat, Wilts.

15th century. Treatise without title on the virtues of herbs.
In Lord Leconfield's library at Petworth House, Sussex.

15th century. Medicinal qualities of herbs.
Phillipps MS. 11077, now in the library of T. Fitzroy Fenwick, Esq., Thirlestaine House, Cheltenham.

II

ENGLISH HERBALS

(Printed books)

The Herbals are listed according to authors, or, in the case of anonymous works, according to the names by which they are usually known, and all known editions are given. In cases where only one copy of an edition is known the library where it is to be found is indicated. Editions mentioned in Ames, Hazlitt, etc., but of which no copies are now known, are listed, but in each case the fact that the only mention of them is to be found in one of the above is stated. [ ] indicates books which are not strictly herbals, but whose omission would make any bibliography of herbals incomplete.

Bartholomæus Anglicus.

1495. [Bartholomæus Anglicus. De proprietatibus rerum.] The seventeenth book of the above—containing nineteen chapters—is on herbs. It was the first original work on plants by an English writer to be printed, and the woodcut at the beginning of the book was probably the first botanical illustration to be printed in an English book.

There is the following note on a slip in the copy of this edition in the British Museum. "This is generally considered to be the finest copy known of a work which is certainly the *chef d'œuvre* of Winkin de Worde's press. The paper on which it is printed is said to be the first ever made in England for the press. See Douce, ii. 278. Dibdin, *Typt. Ant.* ii. 310."

1535. Bartholomeus de Proprietatibus Rerum. Londini in Aedibus Thomæ Berthelete. Regii Impressoris.

1582. Batman uppon Barthōlome His Booke De Proprietatibus Rerum. Newly corrected, enlarged and amended: with such Additions as are requisite unto every seuerall Booke: Taken foorth of the most approved Authors, the like heretofore not translated in English. Profitable for all Estates as well for the benefite of the Mind as the Bodie. London. Imprinted by Thomas East, dwelling by Paules Wharfe.

(For foreign editions, French, Dutch and Spanish translations, see Bibliography of Foreign Printed Herbals, p. 225.)

Banckes's Herbal.

1525. ¶ Here begynnyth a new mater / the whiche sheweth and | treateth of yᵉ vertues & proprytes of her- | bes / the whiche is called | an Herball | ·.· | ¶ Cum gratia & priuilegio | a rege indulto |

Colophon. ¶ Imprynted by me Rycharde Banckes / dwellynge in | Lōdō / a lytel fro yᵉ Stockes in ye Pultry / yᵉ xxv day of | Marche. The yere of our Lorde, M.CCCCC. & XXV. Black-letter 4to.

1526. Second edition of above. Only known copy is in the Cambridge University Library. Title and colophon identical except for slight differences in spelling.

¶ Here begynneth a newe marer / yᵉ whiche sheweth and | treateth of the vertues & propertes of her- | bes / the whiche is callyd | an Herball | ·.· | ¶ Cum priuilegio. |

Colophon. ¶ Imprynted by me Rycharde Banckes / dwellynge in | Lōdō / a lytell fro yᵉ Stockes in yᵉ Pultry / ye xxv daye of June. The yere of our Lorde, M.CCCCC. & XXVI. Black-letter 4to.

1530. (approximate date assigned in the catalogue of the British Museum). A boke of | the propertyes | of herbes the | whiche is | called an | Herbal | ✠ |

Colophon. Imprynted at | London in Fletestrete at | the sygne of the George by | me Robert Red- | man . ·. | ✠ | Black-letter 8vo.

1532–1537 (approximate date assigned by Mr. H. M. Barlow). "'A boke of the propertyes of herbes the which is called an Herbal.' Contains k⁴. 'At the end, Imprynted at London by me John Skot dwellynge in Fauster Lane.' This over his device which is his cypher on a shield, hung on a rose-tree, flowering above the shield, supported by two griffins: at the bottom is a dog nearly couchant; I. S., the initials of his name, one on each side of the trunk of the tree. In the collection of Mr. Alchorne. Twelves."

The above is quoted from Herbert's edition of Ames, 1785. No copy of the work can now be found in any of the chief British libraries.

Mr. Gordon Duff in his list of books printed by John Skot mentions "The Book of Herbes. 12 mo. undated."

The following editions printed by Robert Wyer are all undated. The dates assigned in the British Museum Catalogue are 1530, 1535, 1540.

¶ A newe Her- | ball of Macer, | Translated | out of La- | ten in to | Englysshe.

Colophon. ¶ Imprynted by | me Robert wyer, | dwellynge in saint Martyns pa | ryshe, at the sygne of saynt | Johñ Euangelyst | besyde Charyn | ge Crosse. | ✠ | Secretary type, 8vo.

¶ Hereafter folo | weth the know- | ledge, proper | ties, and the | vertues of | Herbes.

Colophon. ¶ Imprynted by | me Robert Wyer, | dwellynge in saynt Martyns pa- | rysshe, at the sygne of saynt | Johñ Euangelyst, | besyde Charyn | ge Crosse. | ✠ | Secretary type 8vo.

Macers | Herbal | Practy- | syd by | Doctor | Lynacro | Translated out of laten, | into Englysshe, which | shewynge theyr Ope- | raycions & Vertues, | set in the margent | of this Boke, to | the extent you | myght knowe | theyr Ver- | tues.

Colophon. Imprynted by | me Robert wyer | dwellynge in seynt Martyns Pa- | rysshe at the sygne of seynt | Iohñ Euangelyst, besyde Charyn- | ge Crosse. Black-letter 8vo.

The only known copies of the two following editions are in the Bodleian Library.

1541. A boke of | the propertyes | of herbes the whiche | is called an Har | bal, MD. | XLI. |

Colophon. ¶ Imprynted at London | in Paules Churchyearde, | at the Sygne of the may- | dens head by Tho- | mas Petyt. | M.D.X.(I.) Black-letter 8vo.

1546. A boke of | the propertyes | of herbes the | whiche is | called an | Herbal.

Colophon. Imprinted | at London in Fletstrete | at the sygne of the George | nexte to seynt Dunstones churche | by me Wyllyam Myddylton | in the yere of our Lorde | M.CCCCC.XLVI. | The thyrde day | of July | Black-letter 8vo.

1548 (date assigned in the catalogue of the library of the Manchester Medical Society. Only known copy.) ¶ A boke of | the propertes | of herbes the | which is cal | led an her | bal. | ✠ |

Colophon. Imprynted at | London by | Johan Waley, | dwellynge in | Foster Lane. | Black-letter 8vo.

1550 (date assigned in the British Museum Catalogue). A boke of the | properties of Herbes called an her- | ball, whereunto is added the tyme yᵉ | herbes, floures and Sedes shoulde | be gathered to be kept the whole ye- | re, with the vertue of yᵉ Herbes whē | they are stylled. Also a generall rule | of al manner of Herbes drawen out | of an auncient boke | of Physycke by | W. C. |

Colophon. Imprinted at London by Wyllyam | Copland. | Black-letter 8vo.

1552 (date assigned in the British Museum Catalogue). A boke of the | propreties of Herbes called an her | ball, whereunto is added the time yᵉ |

herbes, floures and Sedes shold | be gathered to be kept the whole | yere | wyth the vertue of ye Her- | bes when they are stilled. Al- | so a generall rule of all ma- | ner of Herbes drawen | out of an auncyent | booke of Phisyck | by W. C. |

Colophon. ¶ Imprynted at London in the | Flete strete at the sygne of | the Rose Garland by | me Wyllyam Copland. | for John Wyght |. Black-letter 8vo.

The two following editions published by Anthony Kitson and Richard Kele may be ascribed to Copland's press. No copies exist in the chief British libraries. The titles are quoted from Ames.

" A booke of the properties of Herbes, called an Herball. Whereunto is added the tyme that Herbes, Floures and Seedes should bee gathered to bee kept the whole yeare, wyth the vertue of the Herbes when they are stylled. Also a generall rule of all maner of Herbs, drawen out of an auncient booke of Physicke by W. C. *Walter Carey.* Contains besides X[4] in eights, For him."

1550 (date assigned by Mr. Gordon Duff, but in Ames 1552). " The book of the properties of herbes, called an herball, etc., drawn out of an ancient book of phisyck by W. C."

1550. A lytel | herball of the | properties of her- | bes newely amended & corrected, | with certayne addicions at the ende | of the boke, declaryng what herbes | hath influence of certaine Sterres | and constellations, wherby may be | chosen the beast & most luckye | tymes and dayes of their mini- | stracion, accordynge to the | Moone being in the sig- | nes of heauen, the | which is dayly | appoynted | in the Almanacke, made & gathered | in the yere of our Lorde god | M.D.L. the xii day of Fe- | bruary by Anthonye | Askham Phi- | sycyon.

Colophon. Imprinted at | London in Flete- | strete at the signe of the George | nexte to Saynte Dunstones | Churche by Wylly- | am Powell. In the yeare of oure Lorde | M.D.L. the twelfe day of Marche. Black-letter 8vo.

1550 A litle Her- | ball of the properties of Herbes, | newly amended & corrected, wyth | certayne Additions at the ende of | the boke, declaring what Herbes | hath influence of certain Sterres | and constellations, whereby maye | be chosen the best & most lucky | tymes & dayes of their mini- | stracion, according to the Moone | beyng in the signes of heauē | the which is daily appoī | ted in the Almanacke, | made and gathe- | red in the yeare | of our Lorde | God. | M.D.L. the xii daye of Febru | ary by Anthony Askhā | Physycyon |

Colophon. Imprynted at London, in | Paule's churchyarde, at the signe of the Swanne, by | Ihon Kynge. | Black-letter 8vo.

1555–1561 (approximate date assigned by Mr. H. M. Barlow). ¶ A boke of the | propreties of Herbes called an her | ball, whereunto is added the time y^e^ | herbes, floures and Sedes shold | be gathered to be kept the whole | yere, with the vertue of ye Her | bes when they are stilled. Al- | so a general rule of al ma- | ner of Herbes drawen out of an auncient | boke of Phisyck | by W. C. |

Colophon. ¶ Imprinted at London by | Iohn kynge, for | Abraham Wely |. Black-letter 8vo.

1516. The Grete Herball. Imprented at London in Southwark by me Peter Treveris. MD XVI. the xx day of June.

(Mentioned by Ames. No copy of this edition in any of the chief British libraries and no other record of it.)

1525(?). The Grete herball, which is translated out ye Frensshe into Englysshe. With the Mark of Peter Treveris. Undated.

(Mentioned by Hazlitt, who ascribes the date 1525–6. There is no other record of this edition.)

1527. The grete herball. MDXXVII. 18 April.

(Mentioned by Ames as having been printed by Treveris for Laurence Andrew. No copy of this edition in any of the chief British libraries and no other record of it.)

1526. The grete herball | whiche geueth parfyt knowlege and under- | standyng of all maner of herbes & there gracyous vertues whiche god hath | ordeyned for our prosperous welfare and helth, for they hele & cure all maner | of dyseases and sekenesses that fall or mysfortune to all maner of creatoures | of god created, practysed by many expert and wyse maysters, as Auicenna and | other &c. Also it geueth full parfyte understandynge of the booke lately pryn | ted by me (Peter treueris) named the noble experiens of the vertuous hand | warke of Surgery.

Colophon. ¶ Imprentyd at London in South- | warke by me peter Treueris, dwel- | lynge in the sygne of the wodows. | In the yere of our Lorde god. M.D. | XXVI. the xxvii day of July. Black-letter folio.

1529. Second edition of the above also printed by Treveris. Wording of the title is the same.

Colophon differs from the first edition in that it does not contain the printer's address.

¶ Imprynted at London in South | warke by me Peter Treueris. In | the yere of our Lorde god. M.D.XXIX. | the xvii day of Marce. Black-letter folio.

1539. The great herball | newly corrected. | The contents of this boke. | A table after the Latyn names of all | herbes, | A table after the Englyshe names of all | herbes. | The propertees and qualytes of all | thynges in this booke, | The descrypcyon of urynes, how a man | shall haue trewe know-ledge of all seke- | nesses. | An exposycyon of the wordes obscure and | not well knowen. | A table, quyckly to fynde Remedyes | for all dyseases. | God saue the Kynge. | Londine in Edibus Thome Gybson. | Anno | M.D.XXXIX. Black-letter folio.

This edition contains no cuts.

1550. Edition of "The Grete Herball" mentioned in Ames and Pulteney. No copy of this edition in any of the chief British libraries.

1561. The greate Herball, which | geueth parfyte knowledge & un- | der-standinge of al maner of her | bes, and theyr gracious vertues, whiche God hath ordeyned for | our prosperous welfare and health, for they heale and cure all ma- | ner of diseases and sekenesses, that fall or mysfortune too all | maner of creatures of God created, practysed by many | experte and wyse maysters, as Auicenna, Pandecta, | and more other, &c.

¶ Newlye corrected and dili | gently ouersene. | In the yeare of our Lord | God. M.CCCCC.LXI.

Colophon. Imprynted at London in | Paules churcheyarde, at the signe of the Swane, | by Jhon Kynge. In the yeare of our | Lorde God. M.D.LXI. Black-letter folio.

" *The vertuose boke Of Distyllacyon of the waters of all maner of Herbes.*"

1527. [The vertuose boke of Distyllacyon of the waters of all maner of Herbes / with the figures of the styllatoryes / Fyrst made and compyled by the thyrte yeres study and labour of the most conynge and famous mayster of phisyke / Master Jherom bruynswyke And now newly Translate out of Duyche into Englysshe. Not only to the synguler helpe and profyte of the Surgyens / Physycyens / and Pothecaryes / But also of all maner of people / Parfytely and in dewe tyme and ordre to lerne to dystyll all maner of Herbes / To the Profyte / cure / and Remedy of all maner dysseases and Infirmytees Apparant and not apparant. ¶ And ye shall understand that the waters be better than the Herbes / as Auicenna testefyeth in his fourthe Conon saynge that all maner medicynes ysed with theyr substance / febleth and maketh aged / and weke.

¶ Cum gratia et preuilegio regali.

Colophon. Imprinted at London in the flete strete by me Laurens Andrewe / in the sygne of the golden Crosse. In the yere of our lorde M.CCCC.XXVII (*sic*) the xvii daye of Apryll.

Goddis grace shall euer endure.

Second edition. Title identical with above.

Colophon. Imprynted at London in the flete strete by me Laurens Andrewe / in the Sygne of the golden Crosse. In the yere of our Lorde MCCCCCXXVII, the xviii daye of Apryll.

¶ Goddys grace shall euer endure.]

(This edition, although professedly printed one day later, varies considerably from the preceding.)

William Turner.

1538. [Libellus de | re herbaria novus | in quo herbarum aliquot no- | mina greca, latina & Anglica | habes, vna cum nomini- | bus officinarum, in | gratiam stu- | diose | iuuentutis nunc pri- | mum in lucem | æditus. Londini apud Ioannem Bydellum | Anno dñi. 1538.]

1877. [Libellus de re herbaria novus by William Turner, originally published in 1538. Reprinted in facsimile, with notes, modern names, and A Life of the author, by Benjamin Daydon Jackson, F.L.S. Privately printed. London, 1877.]

1544. Historia de Naturis Herbarum Scholiis et Notis Vallata. Printed at Cologne.

(This book is mentioned by Bumaldus, but is not otherwise known.)

1548. The na | mes of herbes in | Greke, Latin, Englishe, | Duche, and Frenche wyth | the commune names | that Herbaries | and Apoteca | ries use, | Gathered by Wil- | liam Tur | ner.

Colophon. Imprinted | at London by John Day | and Wyllyam Setes, dwel- | lynge in Sepulchres Parish | at the signe of the Resur- | rection a litle aboue Hol- | bourne Conduite. | Cum gratia & priuilegio | ad imprimendum solum.

1881. The names of Herbes by William Turner. A.D. 1548. Edited (with an introduction, an index of English names, and an identification of the plants enumerated by Turner) by James Britten, F.L.S. London. Published for the English Dialect Society, by N. Trübner & Co.

1551. A new Her- | ball, wherein are conteyned the names of Herbes in Greke, La- | tin, Englysh, Duch, Frenche, and | in the Potecaries and Herbari- | es Latin, with the properties | degrees and naturall places of | the same, gathered & made | by Wylliam Turner, | Phisicion unto the | Duke of So- | mersettes | Grace. | Imprinted | at London by Steven | Mierdman. | Anno 1551. | Cum Priuilegio ad imprimendum solum. | And they are to be sold in Paules Churchyarde.

Colophon. Imprinted at London, By Steuen Myerdman, and they are to be soolde in Paules | churchyarde at the sygne of the sprede Egle by | John Gybken.

1562. The seconde parte of Vui- | liam Turners herball, wherein are conteyned the | names of herbes in Greke, Latin, Duche, Frenche, and in the | Apothecaries Latin, and somtyme in Italiane, wyth the ver- | tues of the same herbes | with diuerse confutationes of no small errours, that men of no small learning haue committed in the intreatinge of herbes | of late yeares |

Imprinted at Collen by Arnold Birckman | In the yeare of our Lorde M.D. LXII. | Cum gratia et Priuilegio Reg. Maiest.

1568. The first and seconde partes of the Herbal of William Turner Doctor in Phisick lately ouersene corrected and enlarged with the Thirde parte / lately gathered / and nowe set oute with the names of the herbes / in Greke Latin / English / Duche / Frenche / and in the Apothecaries and Herbaries Latin / with the properties / degrees / and naturall places of the same.

God saue the Quene.

Imprinted at Collen by Arnold Birckman / In the yeare of our Lorde M.D. LXVIII.

Albertus Magnus.

1560 (?). [The boke | of secretes of Albartus Mag | nus, of the vertues of | Herbes, stones and certaine beastes. | Also a boke of the same au | thor, of the marvaylous thin | ges of the world: and of | certaine effectes, cau | sed of certayne | beastes.]

Williyam Bullein.

1562. ¶ BVLLEINS | Bulwarke of defēce | againste all Sicknes, Sornes, and woundes, that dooe | daily assaulte mankinde, whiche Bulwarke is | kepte with Hillarius the Gardiner, Health the | Phisician, with their Chyrurgian, to helpe the | wounded soldiors. Gathered and pra- | ctised frō the moste worthie learn- | ned, bothe old and newe: to | the greate comforte of | mankinde: Doen | by Williyam | Bulleyn, | and ended this Marche, | Anno Salutis. 1562 | ¶ Imprinted at London, by Jhon Kyngston.

1579. BVLLEINS | Bulwarke of Defence against | all Sicknesse, Soarenesse | and VVoundes that | doe dayly assaulte mankinde: Which Bulwarke is | kept with Hilarius the Gardener, and Health | the Phisicion, with the Chirurgian, to helpe the | Wounded Souldiours. Gathered and practised from | the most worthy learned, both olde and new: | to the great comfort of Mankinde: by | VVilliam Bullein, Doctor of Phi- | sicke. 1562. Imprinted | At London by Thomas Marshe, dwellinge | in Fleete streete neare unto Saincte | Dunstanes Chur (*sic*) | 1579. | Eccle. 38. Altissimus creauit de terra medicinam, & vir prudens non abhorrebit illam.

John Maplet.

1567. A greene Forest, or a naturall Historie, Wherein may bee seene first the most sufferaigne Vertues in all the whole kinde of Stones & Mettals: next of Plants, as of Herbes, Trees, & Shrubs, Lastly of Brute Beastes, Foules, Fishes, creeping wormes & Serpents, and that Alphabetically: so that a Table shall not neede. Compiled by John Maplet M. of Arte, and student in Cambridge: extending hereby y[t] God might especially be glorified: and the people furdered. Anno 1567. Imprinted at London by Henry Denham, dwelling in Pater-noster Rovve at the Starre. Anno Domini. 1567. June 3. Cum Priuilegio.

(The dedicatory epistle is to the Earl of Sussex, " Justice of the Forrestes & Chases from Trent Southward; and Captaine of the Gentlemen Pensioners, of the house of the Queene our Soueraigne Ladie, Eliz.").

Pierre Pena and Matthias de l'Obel.

1571. Stirpium Adversaria Nova, | perfacilis vestigatio, luculentaque accessio ad Priscorum, presertim | Dioscoridis et recentiorum, Materiam Medicam. | Quibus propediem accedat altera pars. | Qua | Coniectaneorum de plantis appendix, | De succis medicatis et Metallicis sectio, | Antiquæ e[t] nouatæ Medicine lectiorum remediorū | thesaurus opulentissimus, | De Succedaneis libellus, continentur. | Authoribus Petro Pena & Mathia de Lobel, Medicis. |

Colophon Londini, 1571 | Calendis Januariis excudebat prelum Tho- | mæ Purfœtii ad Lucretie symbolum. | Cum gratia Priuilegii. |

1605. Petrus Pena & Matthias de L'Obel. Dilvcidæ simplicivm medicamenorvm explicationes, & stirpivm adversaria, perfacilis vestigatio, luculentaque accessio ad priscorum, præsertim Dioscoridis & recentiorum materiæ medicæ solidam cognitionem. Londini 1605.

1654. Matthiæ de l'Obel M.D. Botanographi Regii eximii Stirpium Illustrationes. Plurimas elaborantes inauditas plantas, subreptitiis Joh: Parkinsoni rapsodiis ex codice MS insalutato sparsim gravatæ Ejusdem adjecta sunt ad calcem Theatri Botanici Accurante Guil: How, Anglo. Londini Typis Tho: Warren, Impensis Jos: Kirton, Bibliopolæ, in Cæmeterio D. Pauli. 1654.

John Frampton.

1577. Ioyfull | Nevves ovt of | the newe founde worlde, wherein is | declared the rare and singular vertues of diuerse | and sundrie Hearbes, Trees, Oyles, Plantes, and Stones, with | their applications, as well for Phisicke as Chirurgerie, the saied be- | yng well applied bryngeth suche present remedie for | all deseases, as maie seme altogether incredible : | notwithstandyng by practise founde out, | to bee true : Also the portrature of the saied Hearbes, very apt- | ly discribed : Engli- | shed by Jhon | Framp- | ton | Marchaunt |

¶ Imprinted at London in | Poules Churche-yarde, by | Willyam Norton. | Anno Domini. | 1577 |.

1580. Second edition.

1577. The Three | Bookes written in the | Spanishe tonge, by the famous | Phisition D. Monardes, residēt in the | Citie of Seuill in Spaine and | translated into Englishe by | Jhon Frampton | Marchant |

¶ Imprinted at London in | Poules Churche-yarde, by | Willyam Norton. | 1577 |.

(A duplicate of the preceding with a different title-page.)

1596. Ioyfull newes | out of the new-found | worlde | Wherein are declared the rare and | singuler vertues of diuers Herbs, Trees, | Plantes, Oyles & Stones, with their ap- | plications, as well to the vse of phisicts, as of | chirurgery, which being well applyed bring | a present remedie for al diseases, et may | seeme altogether incredible : Notwith- | standing by practice found out | to be true. | Also the portrature of the said Hearbs | very aptlie described : | by John Frampton, Marchant | Newly corrected as by conference with | the olde copies may appeare. Wher- | vnto are added three other bookes | treating of the Bezaar-stone, the herb | Escuerconera, the properties of Iron | and Steele in medicine and the be- | nefit of snow. Printed by E. Allde by the assigne of | Bonham Norton | 1596.

(For the Spanish original and Latin, Italian, French, Flemish and German translations see Bibliography of Foreign Herbals.)

Henry Lyte.

1578. A Niewe Herball | or Historie of Plantes : | wherein is contayned | the whole discourse and perfect description of all sortes of Herbes | and Plantes : their diuers & sundry kindes : | their straunge Figures, Fashions, and Shapes : | their Names, Natures, Operations, and Ver- | tues : and that not onely of those whiche are | here growyng in this our Countrie of | Englande, but of all others also of forrayne Realmes, commonly | used in Physicke. | First set foorth in the Doutche or Almaigne | tongue, by that learned D. Rembert Do- | doens Physition to the Emperour : | And nowe first translated out of French into English, by Hen- | ry Lyte Esquyer. | At London | by me Gerard Dewes, dwelling in | Paules Churchyarde at the signe | of the Swanne. | 1578.

Colophon. Imprinted at Antwerpe, by me | Henry Loë Bookeprinter, and are to be | solde at London in Powels Churchyarde, | by Gerard Dewes.

1586. A New Herball or Historie of Plants: Wherein is contained the whole discourse and perfect description of all sorts of Herbes and Plants: their diuers and sundrie kindes: their Names, Natures, Operations & Vertues: and that not onely of those which are heere growing in this our Countrie of England, but of all others also of forraine Realms commonly used in Physicke. First set foorth in the Dutch or Almaigne toong by that learned D. Rembert Dodoens Physition to the Emperor: And now first translated out of French into English by Henrie Lyte Esquier. Imprinted at London by Ninian Newton. 1586.

1595. Title identical with above, except for the addition of "Corrected and Amended. Imprinted at London by Edm: Bollifant, 1595."

1619. A New Herbal or Historie of Plants: Wherein is contained the whole discourse and perfect description of all sorts of Herbes and Plants: their diuers and sundry kindes their Names, Natures, Operations and Vertues: and that not onely of those which are here growing in this our Country of England but of all others also of forraine Realmes commonly used in Physicke. First set forth in the Dutch or Almaigne tongue by the learned D. Rembert Dodoens Physicion to the Emperor; and now first translated out of French into English by Henry Lyte Esquire. Corrected and Amended. Imprinted at London by Edward Griffin. 1619.

William Ram.

1606. Rams little Dodoen. A briefe Epitome of the New Herbal, or History of Plants. Wherein is contayned the disposition and true declaration of the Phisike helpes of all sortes of herbes and Plants, under their names and operations, not onely of those which are here in this our Countrey of England growing but of all others also of other Realmes, Countreyes and Nations used in Phisike: Collected out of the most exquisite newe Herball, or History of Plants first set forth in the Dutch or Almayne tongue by the learned and worthy man of famous memory, D. Rembert Dodeon, (*sic*) Phisicion to the Emperour; And lately translated into English by Henry Lyte, Esquire; And now collected and abbridged by William Ram, Gent. Pandit Oliua suos Ramos.

Imprinted at London by Simon Stafford, dwelling in the Cloth Fayre, at the signe of the three Crownes. 1606.

William Langham.

1579. The Garden of Health: containing the sundry rare and hidden vertues and properties of all kindes of Simples and Plants. Together with the manner how they are to bee used and applyed in medicine for the health of mans body, against diuers diseases and infirmities most common amongst men. Gathered by the long experience and industry of William Langham, Practitioner in Physicke. London. Printed by Thomas Harper with permission of the Company of Stationers.

1633. Second edition. Identical title with the addition "The Second edition corrected and amended."

Thomas Newton.

1587. An | Herbal For | the Bible. | Containing A Plaine | and familiar exposition | of such Similitudes, Parables, and | Metaphors, both in the olde Testament and | the Newe, as are borrowed and taken from | Herbs, Plants, Trees, Fruits, and Simples, | by obseruation of their Vertues, qualities, natures, proper- | ties, operations, | and effects : | And | by the Holie Pro- | phets, Sacred Writers, | Christ himselfe, and his blessed Apostles | usually alledged, and unto their heauenly | Oracles, for the better beautifieng | and plainer opening of | the same, profitably | inserted | Drawen into English by Thomas | Newton. | Imprinted at London by Ed- | mund Bollifant | 1587 |
(The dedicatory epistle is to the Earl of Essex.)

John Gerard.

1596. [Catalogus arborum fruticum ac plantarum tam indigenarum quam exoticarum, in horto Ioannis Gerardi civis et Chirurgi Londinensis nascentium-Londini. Ex officina Roberti Robinson 1596.]
1599. Second edition. Londini. Ex officina Arnoldi Hatfield, impensis Ioannis Norton. (The only known copy of the first edition is in the Sloane collection in the British Museum.)
1876. Modern reprint with notes, etc., by B. D. Jackson.
1597. The | Herball | or Generall | Historie of | Plantes. | Gathered by John Gerarde | of London Master in | Chirurgerie. | Imprinted at London by | John Norton. | 1597.
Colophon. Imprinted at London by Edm Bollifant, | for Bonham & John | Norton M.D.XCVII.
1633. The | Herball | or Generall | Historie of | Plantes. | Gathered by John Gerarde | of London Master in | Chirurgerie | Very much Enlarged and Amended by |Thomas Johnson | Citizen and Apothecarye | of London.
1636. Second edition of the above.

John Parkinson.

1629. [Paradisi in Sole Paradisus Terrestris. A Garden of all Sorts of Pleasant Flowers Which Our English Ayre will Permitt to be noursed up: with A Kitchen garden of all manner of herbes, rootes, & fruites, for meate or sause used with us, and An Orchard of all sorte of fruit bearing Trees and shrubbes fit for our Land together With the right orderinge planting & preseruing of them and their uses & vertues. Collected by John Parkinson Apothecary of London 1629.
Colophon. London. Printed by Humfrey Lownes and Robert Young at the signe of the Starre on Bread-Street hill. 1629.
1656. Second edition. Title, etc., identical with above.
1904. Facsimile reprint. Paradisi in Sole. Paradisus Terrestris by John Parkinson. Faithfully reprinted from the edition of 1629. Methuen & Co.]

1640. Theatrum Bo | tanicum : | The Theater of Plants | or a Herball of | a | large extent : | containing therein a more ample and | exact History and declaration of the Physicall Herbs | and Plants that are in other Authors, encreased by the accesse of | many hundreds of newe, | rare and strange Plants from all parts of | the world, with sundry Gummes and other Physicall Materi | als than hath been hitherto published by any before, and | a most large demonstration of their Names and Vertues. | Shewing withall the many errors and differences & | oversights of Sundry Authors that have formerly written of | them, and a certaine confidence, or most probable con | jecture of the true and Genuine Herbes | and Plants. | Distributed into Sundry Classes or Tribes for the | more easie knowledge of the many Herbes of one nature | and property with the chief notes of Dr. Lobel, Dr. Bonham | and others inserted therein. | Collected by the many yeares travaile, industry and experience in this subject, by John Parkinson Apothecary of London, and the King's Herbarist. And Published by the King's Majestyes especiall priviledge. London. Printed by Tho. Cotes. 1640.

Leonard Sowerby.

1651. [The Ladies Dispensatory, containing the Natures, Vertues, and Qualities of all Herbs, and Simples usefull in Physick. Reduced into a Methodicall Order, for their more ready use in any sicknesse or other accident of the Body. The like never published in English. With An Alphabeticall Table of all the Vertues of each Herb, and Simple. London. Printed for R. Ibbitson, to be sold by George Calvert at the Halfe-Moon in Watling Street. 1651.]

Robert Pemell.

1652. [Tractatus, De facultatibus Simplicium, A Treatise of the Nature and Qualities of such Simples as are most frequently used in Medicines. Methodically handled for the benefit of those that understand not the Latine Tongue. By Robert Pemell, Practitioner of Physick, at Cranebrooke in Kent. London, Printed by M. Simmons, for Philemon Stephens, at the guilded Lyon in St. Pauls Church-yard. 1652.

1653. Second Part of the above " Treatise." London, Printed by J. Legatt, for Philemon Stephens, at the guilded Lion in Paul's Church-yard. 1653.]

Nicholas Culpeper.

1652. The English Physician Or an Astrologo-physical Discourse of the Vulgar Herbs of this Nation Being a Compleat Method of Physick whereby a man may preserve his Body in health; or cure himself, being sick, for three pence charge, with such things one-ly as grow in England, they being most fit for English Bodies.

Herein is also shewed,

1. The way of making Plaisters, Oyntments, Oyls, Pultisses, Syrups, Decoctions, Julips, or Waters of all Sorts of Physical Herbs, that you may have them ready for your use at all times of the year.

2. What Planet governeth Every Herb or Tree used in Physick that groweth in England.

3. The Time of gathering all Herbs, but vulgarly and astrologically.

4. The way of drying and Keeping the Herbs all the year.

5. The way of Keeping the Juyces ready for use at all times.

6. The way of making and keeping all Kinde of usefull Compounds made of Herbs.

7. The way of mixing Medicines according to Cause and Mixture of the Disease, and Part of the Body afflicted.

By N. Culpeper, Student in Physick and Astrology.

London, Printed for the benefit of the Common-wealth of England. 1652.

(This is the edition repudiated by the author in subsequent editions as incorrect and unauthorised.)

Subsequent editions 1653, 1661, 1693, 1695, 1714, 1725, 1733, 1784, 1792, 1814, 1820.

1818. (*Welsh translation.*) Herbal, Neu Lysieu-Lyfr. Y Rhan Gyntaf, Yn Cynnwys Go o Gynghorion Teuluaidd Hawdd iw cael; Wedi ei casglu allan o Waith. N. Culpeper. Ag amrywiol eraill, a'r rhan fwyaf o honynt wedi eu profi yn rhinwellol ac effeilhiol i symud yr amrywrol ddoluriau ac y mae ein Cyrph llygredig yn ddarostyngedig iddynt: Ac y maent yn hollawl ilw defnyddw o Ddail a Llysiau ein bwlad ein hunain. Cewch hefyd gyfar wyddyd i ollwng Gwaed, ac y gymeryd Purge. Yr ail argraphiad. Gan D. T. Jones. Caernarfon, Argraphwyd Gan L. E. Jones. 1818.

1862. Second edition of the above.

William Coles.

1656. The Art of Simpling. An Introduction to the Knowledge and Gathering of Plants. Wherein the Definitions, Divisions, Places, Descriptions, Differences, Names, Vertues, Times of flourishing and gathering, Uses, Temperatures, Signatures and Appropriations of Plants, are methodically laid down. Whereunto is added A Discovery of the Lesser World. By W. Coles. London. Printed by J. G. for Nath: Brook at the Angell in Cornhill. 1656.

1657. Adam in Eden, or Nature's Paradise. The History of Plants, Fruits, Herbs and Flowers. With their several Names, whether Greek, Latin, or English; the places where they grow; their Descriptions and Kinds; their times of flourishing and decreasing; as also their several Signatures, Anatomical Appropriations, and particular Physical Vertues; together with necessary Observations on the Seasons of Planting, and gathering of our English Simples, with Directions how to preserve them in their Compositions or otherwise. A Work of such a Refined and Useful Method that the Arts of Physick and Chirurgerie are so clearly Laid Open, that

Apothecaries, Chirurgions, and all other ingenuous Practitioners, may from our own Fields and Gardens, best agreeing with our English Bodies, on emergent and Sudden occasions, compleatly furnish themselves with cheap, easie, and wholesome Cures for any part of the body that is ill-affected. For the Herbarists greater benefit, there is annexed a Latin and English Table of the several names of Simples; with another more particular Table of the Diseases, and their Cures, treated of in this so necessary a Work. By William Coles, Herbarist. Printed by J. Streater for Nathaniel Brooke.

Robert Lovell.

1659. ΠΑΜΒΟΤΑΝΟΛΟΓΙΑ. Sive Enchiridion Botanicum. Or a compleat Herball, Containing the Summe of what hath hitherto been published either by Ancient or Moderne Authors both Galenicall and Chymicall, touching Trees, Shrubs, Plants, Fruits, Flowers, etc. In an Alphabeticall order: wherein all that are not in the Physick Garden in Oxford are noted with asterisks. Shewing their Place, Time, Names, Kindes, Temperature, Vertues, Use, Dose, Danger and Antidotes. Together with an Introduction to Herbarisme, etc. Appendix of Exoticks. Universall Index of plants: shewing what grow wild in England. By Robert Lovell. Oxford. Printed by William Hall for Ric Davis. An. 1659.

1665. Second edition. ΠΑΜΒΟΤΑΝΟΛΟΓΙΑ. Sive Enchiridion Botanicum. Or a compleat Herball, Containing the Summe of Ancient and Moderne Authors, both Galenical and Chymical, touching Trees, Shrubs, Plants, Fruits, Flowers, etc. In an Alphabetical order: wherein all that are not in the Physick Garden in Oxford are noted with Asterisks. Shewing their Place, Time, Names, Kinds, Temperature, Vertues, Use, Dose, Danger and Antidotes. Together with An Introduction to Herbarisme, etc. Appendix of Exoticks. Universal Index of Plants: shewing what grow wild in England. The second Edition with many Additions mentioned at the end of the Preface. By Robert Lovell.
Oxford. Printed by W. H. for Ric. Davis. 1665.

John Josselyn.

1672. [New England's Rarities Discovered in Birds, Beasts, Fishes, Serpents, and Plants of that country. Together with the Physical and Chyrurgical Remedies wherewith the Natives constantly use to Cure their Distempers, Wounds and Sores. Also A perfect Description of an Indian Squa in all her Bravery; with a Poem not improperly Conferr'd upon her. Lastly A Chronological Table of the most remarkable Passages in that Country amongst the English. Illustrated with Cuts. By John Josselyn Gent.
London Printed for G. Widdowes at the Green Dragon in St. Pauls Church-yard, 1672.]

W. Hughes.

1672. The American Physitian; Or a Treatise of the Roots, Plants, Trees, Shrubs, Fruit, Herbs, etc., growing in the English Plantations in America. Describing the Place, Time, Names, Kindes, Temperature, Vertues and Uses of them, either for Diet, Physick, etc. Whereunto is added A Discourse of the Cacao-Nut-Tree, and the use of its Fruit, with all the ways of making Chocolate. The like never extant before. By W. Hughes.
London, Printed by J. C. for William Crook, at the Green Dragon without Temple-Bar. 1672.

John Archer.

1673. A Compendious Herbal, discovering the Physical Vertue of all Herbs in this Kingdom, and what Planet rules each Herb, and how to gather them in their Planetary Hours. Written by John Archer, One of His Majesties Physicians in Ordinary. London, Printed for the Author, and are to be sold at his House at the Sign of the Golden Ball in Winchester Street, near Broad Street. 1673.

Robert Morison.

1680. [Plantarum Historiæ Universalis Oxoniensis. Pars Secunda seu Herbarum Distributio Nova, per Tabulas Cognationis & Affinitatis Ex Libro Naturæ Observata & Detecta. Auctore Roberto Morison. Medico & Professore Botanico Regio, nec non Inclytæ & Celeberrimæ Universitatis Oxoniensis P. B. ejusdemque Hort. Botan. Præfecto primo. Oxonii, E Theatro Sheldoniano Anno Domini M.D.C.LXXX.
1699. Pars tertia. Partem hanc tertiam, post Auctoris mortem, hortatu Academiæ explevit & absolvit Jacobus Bobartius forte præfectus.]
(The first part was never published.)

John Ray.

1686. [Historia Plantarum Species hactenus editas aliasque insuper multas noviter inventas & descriptas complectens. In qua agitur primo De Plantis in genere, Earumque Partibus, Accidentibus & Differentiis; Deinde Genera omnia tum summa tum Subalterna ad Species usque infimas, Notis suis certis & Characteristicis Definita, Methodo Naturæ vestigiis insistente disponuntur; Species Singulæ accurate describuntur, obscura illustrantur, omissa supplentur, superflua resecantur, Synonyma necessaria Adjiciuntur; Vires denique & Usus recepti compendiò traduntur. Auctore Joanne Raio, E Societate Regiâ & S.S. Individuæ Trinitatis Collegii apud Cantabrigienses Quondam Socio.
Londini Mariæ Clark: Prostant apud Henricum Faithorne Regiæ Societatis Typographum ad Insigne Rosæ in Cæmeterio. D Pauli. CIↃ IↃ CLXXXVI.]

Leonard Plukenet.

1690. [Leonardi Plukenetij Phytographia. Sive Stirpium Illustriorum & minus cognitarum Icones, Tabulis Æneis, Summa diligentia elaboratæ, Quarum unaquæg Titulis descriptorijs ex Notis Suis proprijs, & Characteristicis desumptis, insignita; ab alijs ejusdem Sortis facile discriminatur. Pars prior Meminisse juvabit. Londini MDCXC, Sumptibus Autoris.]

William Westmacott.

1694. ΘΕΟΛΟΒΟΤΑΝΟΛΟΓΙΑ. Sive Historia Vegetabilium Sacra: or, a Scripture Herbal; wherein all the Trees, Shrubs, Herbs, Plants, Flowers, Fruits, &c., Both Foreign and Native, that are mentioned in the Holy Bible, (being near Eighty in Number) are in an Alphabetical Order, Rationally Discoursed of, Shewing, Their Names, Kinds, Descriptions, Places, Manner of Propagation, Countries, various Uses, Qualities and Natural Principles, &c. Together with their Medicinal Preparations, Virtues and Dose, Galenically and Chymically handled and Performed according to the newest Doctrines of Philosophy, Herbarism and Physick. The whole being Adorned with variety of Matter, and Observations, not only Medicinall, but Relating to the Alimental and Mechanical Uses of the Plants. Fit for Divines, and all Persons of any other Profession and Calling whatsoever, that use to read the Holy Scriptures, wherein they find not only Physick for the Soul, but also with the help of this Herbal, (may the better understand the Bible, which also yields them) safe Medicines for the Cure of their Corporal Diseases. The like never extant before. By William Westmacott of the Borough of Newcastle under Line, in the County of Stafford, Physican. Adoro Scripturæ Plenitudinem. Tertul. London, Printed for T. Salusbury, at the King's-Arms next St. Dunstan's Church in Fleet-street. 1694.

John Pechey.

1694. The Compleat Herbal of Physical Plants. Containing All such English and Foreign Herbs, Shrubs and Trees, as are used in Physick and Surgery. And to the Virtues of those that are now in use, is added one Receipt, or more, of some Learned Physician. The Doses or Quantities of such as are prescribed by the London Physicians, and others, are proportioned. Also Directions for Making Compound-Waters, Syrups Simple and Compound, Electuaries, Pills, Powders, and other Sorts of Medicines. Moreover, The Gums, Balsams, Oyls, Juices, and the like, which are sold by Apothecaries and Druggists, are added to this Herbal; and their Virtues and Uses are fully described. By John Pechey, Of the College of Physicians, in London. Printed for Henry Bonwicke, at the Red Lyon in St. Paul's Church-yard. 1694.

William Salmon.

1710. Botanologia. The English Herbal: or History of Plants. Containing I. Their Names, Greek, Latine and English. II. Their Species, or various Kinds. III. Their Descriptions. IV. Their Places of Growth. V. Their Times of Flowering and Seeding. VI. Their Qualities or Properties. VII. Their Specifications. VIII. Their Preparations, Galenick and Chymick. IX. Their Virtues and Uses. X. A Complete Florilegium, of all the choice Flowers cultivated by our Florists, interspersed through the whole Work, in their proper Places; where you have their Culture, Choice, Increase, and Way of Management, as well for Profit as for Delectation. Adorned with Exquisite Icons or Figures, of the most considerable Species, representing to the Life, the true Forms of those Several Plants. The whole in an Alphabetical Order. By William Salmon, M.D. London: Printed by I. Dawks, for H. Rhodes, at the Star, the Corner of Bride-Lane, in Fleet-Street; and J. Taylor, at the Ship in Pater-noster-Row. M.DCC.X.

(The dedicatory epistle is to Queen Anne.)

James Petiver.

1715. [Hortus Peruvianus Medicinalis: or, the South-Sea herbal. Containing the names, figures, vse, &c., of divers medicinal plants, lately discovered by Pere L. Feuillee, one of the King of France's herbalists. To which are added, the figures, &c., of divers American gum-trees, dying woods, drugs, as the Jesuits bark-tree and others, much desired and very necessary to be known by all such as now traffick to the South-Seas or reside in those parts.]

(Undated.) Botanicum Londinense, or London Herbal. Giving the Names, Descriptions and Virtues &c. of such Plants about London as have been observed in the several Monthly Herborizings made for the Use of the young Apothecaries and others, Students in the Science of Botany or Knowledge of Plants.

(Undated.) Botanicum Anglicum, or The English Herball: Wherein is contained a curious Collection of Real Plants being the true Patterns of such Trees, Shrubs and Herbs as are observed to grow Wild in England. By which any one may most easily attain to the Speedy and True Knowledge of them. With an Account (affixed to each Plant) of their Names, Places where Growing, and Times of Flourishing: As also what Parts and Preparations, of Each Physical Plant, are most in Use. And for the farther Instruction and Satisfaction of such, who are Lovers of Plants, The Composer of this Collection chose to make his chiefest References to the General History, Catalogue and Synopsis of that Learned Author, and most Judicious Botanist, Mr. John Ray: As also to our Two most Esteemed English Herballs, Johnson upon Gerard and Parkinson; and for your more speedy finding each Plant, he hath quoted the Page, wherein you may observe its Name, Figure or Description.

Sold by Samuel Smith at the Princes-Arms in St. Paul's Church-yard. London.

[Undated. The Virtues of several Sovereign Plants found wild in Maryland with Remarks on them.]

Tournefort's Herbal.

1716. The Compleat Herbal: or, the Botanical Institutions of Mr. Tournefort, Chief Botanist to the late French King. Carefully translated from the original Latin. With large Additions, from Ray, Gerard, Parkinson, and others, the most celebrated Moderns; Containing what is further observable upon the same Subject, together with a full and exact Account of the Physical Virtues and Uses of the several Plants; and a more compleat Dictionary of the Technical Words of this Art, than ever hitherto published: Illustrated with about five hundred Copper Plates, containing above four thousand different Figures, all curiously engraven. A Work highly Instructive, and of general Use.

In the Savoy: Printed by John Nutt, and Sold by J. Morphew near Stationers-Hall, and most Booksellers in Great-Britain and Ireland. 1716.

Joseph Miller.

1722. Botanicum Officinale; or a Compendious Herbal: giving an account of all such Plants as are now used in the Practice of Physick. With their Descriptions and Virtues. By Joseph Miller. London: Printed for E. Bell in Cornhill, J. Senex in Fleet-Street, W. Taylor in Pater-noster-Row, and J. Osborn in Lombard-Street. M.DCC.XXII.

(The book is dedicated to Sir Hans Sloane.)

Patrick Blair.

1723. Pharmaco-Botanologia: or, An Alphabetical and Classical Dissertation on all the British Indigenous and Garden Plants of the New London Dispensatory. In which Their Genera, Species, Characteristick and Distinctive Notes are Methodically described; the Botanical Terms of Art explained; their Virtues, Uses, and Shop-Preparations declared. With many curious and useful Remarks from proper Observation. By Patrick Blair, M.D., of Boston in Lincolnshire and Fellow of the Royal Society. London: Printed for G. Strahan at the Golden Ball over-against the Royal Exchange in Cornhill; W. and J. Innys at the West End of St. Paul's Church-yard; and W. Mears at the Lamb, without Temple Bar. MDCCXXIII.

Elizabeth Blackwell.

1737. A Curious Herbal, Containing Five Hundred Cuts, of the most useful Plants, which are now used in the Practice of Physick. Engraved on folio Copper Plates, after Drawings, taken from the Life. By Elizabeth Blackwell. To which is added a short Description of ye Plants; and their common Uses in Physick. London. Printed for Samuel Harding in St. Martin's Lane. MDCCXXXVII.

1757. Herbarium Blackwellianvm Emendatvm et Anetivm id est Elisabethæ Blackwell Collectio Stirpium Qvæ in Pharmacopoliis ad Medicvm vsvm asservantvr Qvarvm Descriptio et Vires ex Anglico idiomate in Latinvm conversæsistvntvr figuræ maximam partem ad naturale Exemplar emendantvr floris frvctvsqve partivm repræsentatione avgentvr et Probatis Botanicorvm nominibvs cum præfatione Tit. Pl. D.D. Christoph. Iacobi Trew . Excvdit figvras pinxit atqve in æs incidit Nicolavs-Fridericvs Eisenbergervs sereniss . Dvcis Saxo-Hildbvrg . Pictor avlicvs Norimbergæ Degens . Norim bergæ Typis Christiani de Lavnoy Anno MDCCLVII.

Thomas Short.

1747. Medicina Britannica: or a Treatise on such Physical Plants, as are Generally to be found in the Fields or Gardens in Great-Britain: Containing A particular Account of their Nature, Virtues, and Uses. Together with The Observations of the most learned Physicians, as well ancient as modern, communicated to the late ingenious Mr. Ray, and the learned Dr. Sim. Pauli . Adapted more especially to the Occasions of those, whose Condition or Situation of Life deprives them, in a great Measure, of the Helps of the Learned. By Tho. Short of Sheffield, M.D. London. Printed for R. Manby & H. Shute Cox, opposite the Old Baily on Ludgate-Hill. MDCCXLVII.

1748. [A complete History of Drugs. Written in French By Monsieur Pomet, Chief Druggist to the late French King Lewis XIV. To which is added what is farther observable on the same Subject, from Mess Lemery and Tournefort, Divided into Three Classes, Vegetable, Animal, and Mineral; With their Use in Physic, Chemistry, Pharmacy, and several other Arts. Illustrated with above Four Hundred Copper-Cuts, curiously done from the Life; and an Explanation of their different Names, Places of Growth, and Countries where they are produced; with the Methods of distinguishing the Genuine and Perfect, from the Adulterated, Sophisticated and Decayed; together with their Virtues, &c. A Work of very great Use and Curiosity. Done into English from the Originals. London. Printed for J. and J. Bonwicke, S. Birt, W. Parker, C. Hitch, and E. Wicksteed. MDCCXLVIII.]

(The above is dedicated to Sir Hans Sloane.)

James Newton.

1752. A compleat Herbal of the late James Newton, M.D., Containing the Prints and the English Names of several thousand Trees, Plants, Shrubs, Flowers, Exotics, etc. All curiously engraved on Copper-Plates. London: Printed by E. Cave at S. John's Gate; and sold by Mr. Watson, an Apothecary, over-against St. Martin's Church, in the Strand; Mr. Parker, at Oxford; Mr. Sandby, at the Ship, in Fleet-street. M,DCC,LII.

"Sir" John Hill.

1755. The Family Herbal, or an account of all those English Plants, which are remarkable for their virtues, and of the Drugs which are produced by Vegetables of other Countries; with their descriptions and their uses, as proved by experience. Also directions for the gathering and preserving roots, herbs, flowers, and seeds; the various methods of preserving these simples for present use; receipts for making distilled waters, conserves, syrups, electuaries, Juleps, draughts, &c., &c., with necessary cautions in giving them. Intended for the use of families. By Sir John Hill, M.D., F.R.A. of Sciences at Bourdeaux.

Subsequent editions, 1812, 1820.

1756. The British Herbal; An History of Plants and Trees, Natives of Britain, cultivated for use, or raised for beauty. By John Hill, M.D. London. Printed for T. Osborne and J. Shipton, in Grays-Inn; J. Hodges, near London-Bridge; J. Newbery in S. Paul's Church-Yard; B. Collins; And S. Crowder and H. Woodgate, in Pater-noster-Row. MDCCLVI.

1769. Herbarium Britannicum Exhibens Plantas Britanniæ Indigenas secundum Methodum floralem novam digestas. Cum Historia, Descriptione, Characteribus Specificis, Viribus, et Usis. Auctore Johanne Hill, Medicinæ Doctore, Academiæ Imperialis Naturæ Curiosorum Dioscoride quarto, &c. Londini: Sumptibus auctoris. Prostant apud Baldwin, Ridley, Nourse, Becket, Davies, Cambell, Elmsly Bibliopolis. MDCCLXIX.

Timothy Sheldrake.

1759 (*circ.*). Botanicum Medicinale; An Herbal of Medicinal Plants on the College of Physicians List. Describing their Places of Growth, Roots, Bark, Leaves, Buds, Time of Flowering, Blossoms, Flowers, Stiles, Chives, Embrio's, Fruits, Farina, Colours, Seeds, Kernels, Seed-Vessels, Parts used in Medicine, Preparations in the Shops, Medicinal Virtues, Names in Nine Languages. Most beautifully engraved on 120 Large Folio Copper-Plates, From the Exquisite Drawings of the late Ingenious T. Sheldrake. English Plants are drawn from Nature to the greatest Accuracy, Flowers, or Parts, too small to be distinguished, are magnified. Nothing in any Language exceeds this Thirty Years laborious Work, of which may truly be said that Nature only equals it, every Thing of the Kind, hitherto attempted, being trivial, compared to this inimitable Performance. Designed to promote Botanical Knowledge, prevent Mistakes in the Use of Simples in compounding and preparing Medicines, to illustrate, and render such Herbals as want the Just Representations in their proper Figures and Colour more useful. Necessary to such as practise Physic, Pharmacy, Chemistry, &c., entertaining to the Curious, the Divine and Philosopher, in contemplating these wonderful Productions,—useful to Painters, Heralds, Carvers, Designers, Gardeners, etc. The Colours of every part are minutely described; for Utility it must be esteemed to any Hortus Siccus extant. The Means to preserve Fruits, or to dry Flowers, in their Native Form and Colour are not yet discovered; Plants cannot be preserved to Perfection. The Flowers, when coloured, are represented in their original Bloom, and Fruits in the inviting Charms of Maturity. To which is now added His Tables for finding the Heat and Cold in all Climates, that Exotic Plants may be raised in Summer, and preserved in Winter. London. Printed for J. Millan, opposite the Admiralty, Whitehall.

John Edwards.

1770. The British Herbal containing one hundred Plates of the most beautiful and scarce Flowers and useful Medicinal Plants which blow in the open Air of Great Britain, accurately coloured from Nature, with their Botanical Characters, and A short account of their Cultivation, etc., etc. By John Edwards. London: Printed for the Author; and sold by J. Edmonson, Painter to Her Majesty in Warwick Street, Golden Square; and J. Walter at Homer's Head, Charing-Cross. MDCCLXX.

1775. A select Collection of One Hundred Plates; consisting of the most beautiful exotic and British Flowers which blow in our English Gardens, accurately drawn and Coloured from Nature, with their Botanic Characters, and a short account of their Cultivation, Their uses in Medicine, with Their Latin and English Names. By John Edwards. London: Printed for S. Hooper, No. 25 Ludgate-Hill. M.DCC.LXXV.

William Meyrick.

1789. The New Family Herbal; or Domestic Physician: Enumerating with accurate Descriptions, All the known Vegetables which are any way remarkable for medical efficacy; with an account of their Virtues in the Several Diseases incident to the Human Frame. Illustrated with figures of the most remarkable plants, accurately delineated and engraved. By William Meyrick, Surgeon. Birmingham, Printed by Pearson and Rollason, and Sold by R. Baldwin, Pater-noster Row London. MDCCLXXXIX.

1790. Second edition—Title, etc., identical with above.

Henry Barham.

1794. Hortus Americanus: Containing an account of the Trees, Shrubs, and other Vegetable Productions, of South-America and the West India Islands, and particularly of the Island of Jamaica; Interspersed with many curious and useful Observations, respecting their Uses in Medicine, Diet, and Mechanics. By the late Dr. Henry Barham. To which are added a Linnæan Index, etc., etc., etc. Kingston, Jamaica: printed and published by Alexander Arkman, Printer to the King's most Excellent Majesty, and to the Honourable House of Assembly. MDCCXCIV.

Robert John Thornton.

1810. A Family Herbal: a Familiar Account of the Medical Properties of British and Foreign Plants, also their uses in dying, and the various Arts, arranged according to the Linnæan System, and illustrated by two hundred and fifty-eight engravings from plants drawn from Nature by Henderson, and engraved by Bewick of Newcastle. By Robert John Thornton, M.D., Member of the University of Cambridge, and of the Royal London College of Physicians; Lecturer on Botany at Guy's Hospital; Author of a Grammar of Botany, the Philosophy of Medicine, etc. London: Printed for B. & R. Crosby and Co., Stationer's Court, Ludgate Street.
1814. Second edition.

Jonathan Stokes.

1812. A Botanical Materia Medica, Consisting of the Generic and Specific Characters of the Plants used in Medicine and Diet, with Synonyms, And references to Medical authors, By Jonathan Stokes, M.D. In Four volumes. London, Printed for J. Johnson and Co. St. Paul's Churchyard. 1812.

Thomas Green.

1816. The Universal Herbal; or, Botanical, Medical, and Agricultural Dictionary. Containing an account of All the known plants in the World, arranged according to the Linnean System. Specifying the uses to which they are or may be applied, whether as Food, as Medicine, or in the Arts and Manufactures. With the best methods of Propagation, and the most recent agricultural improvements. Collected from indisputable Authorities. Adapted to the use of the Farmer—the Gardener—the Husbandman—the Botanist—the Florist—and Country Housekeepers in General. By Thomas Green. Liverpool. Printed at the Caxton Press by Henry Fisher, Printer in Ordinary to His Majesty. Sold at 87, Bartholomew Close, London.
1824. Second edition.

John Lindley.

1838. Flora Medica; A Botanical Account of all the more important plants used in Medicine, in different parts of the world. By John Lindley, Ph.D., F.R.S., Professor of Botany in University College, London; Vice-Secretary of the Horticultural Society, etc. etc. etc. London: Printed for Longman, Orme, Brown, Green, and Longmans, Paternoster-Row. 1838.

The majority of sixteenth- and early seventeenth-century gardening books devote considerable space to herbs. See especially:—

1563. Thomas Hill. The proffitable Arte of Gardening.
1594. Sir Hugh Platt. The Garden of Eden.
1617. Gervase Markham. The Country Housewife's Garden.
1618. William Lawson. A new Orchard and Garden with the Country Housewife's Garden.

III

FOREIGN HERBALS

(Printed books)

This list includes only the chief works, and those which have some connection with the history of the herbal in England. With the exception of the *Arbolayre*, copies of all the incunabula herbals mentioned below are to be found in the British Museum. Copies in American libraries are noted in the list.

Bartholomæus Anglicus.

1470. Bartholomæus Anglicus. Liber de proprietatibus rerum. Printed at Basle with the type used by both Richel and Wensler.

1470(?) Liber de proprietatibus rerum Bartholomei Anglici. Printed at Cologne by Ulrich Zell.

Subsequent editions, 1480, 1481, 1482, 1483, 1485, 1488, 1491, 1492, 1519, 1601.

(*French translation.*)

(A translation was made by Fr. Jehan Corbichon in 1372 for Charles V. of France.)

1482. Cy commence vng tres excellent liure nomme le proprietaire des choses par Fr. Jehan Corbichon. Printed at Lyons.

Subsequent editions printed at Lyons, 1485, 1491, 1498 (?), 1525, 1530 (?), 1539, 1556.

1485. (*Dutch translation.*) Printed at Haarlem by Jacop Bellaert.

1494. (*Spanish translation.*) El libro de propietatibus (*sic*) rerum trasladado de latin en romance por Vincente de burgos.

1529. Another edition printed at Toledo.

Das pûch der natur.

1475. Konrad von Megenburg. Das pûch der natur. Printed at Augsburg by Hanns Bämler.

(There are a large number of MSS. of the above extant, eighteen of them being in the Vienna library. Eighty-nine herbs and their virtues are described. The woodcuts in this book are exceptionally fine. (There is only one of plants.) In some copies the woodcuts are coloured by a contemporary artist, possibly Bämler himself, for he was well known as an illuminator before he began printing. Though not strictly a herbal, the above is included in this list, as this and *Liber de Proprietatibus Rerum* are the earliest printed books containing a section on herbs.)

1478. Another edition.

1499. Another edition. Printed at Augsburg by Hanns Schönsperger. Cuts are copies of those in the first edition, with the addition of two others from the Strassburg Hortus Sanitatis of *circa* 1490.

Albertus Magnus.

1478. Albertus Magnus. Liber aggregationis seu liber secretorum. Alberti Magni de virtutibus herbarum animalium et mirabilis mundi. (*Colophon*) per Johannem de Annunciata de Augusta.
1500. Edition printed at Rouen by Thomas Laisne.
(This book claims Albertus Magnus as its author, but is wholly unworthy of that great scholar.)

Herbarium Apuleii Platonici.

1480 (*circ.*). Incipit Herbarium Apuleii Platonici ad Marcum Agrippam. Printed at Rome by Philippus de Lignamine, courtier and physician to Sixtus IV. First impression dedicated to Cardinal de Gonzaga. Second impression to Cardinal de Ruvere. The copy in the British Museum has the Ruvere dedication.
America: Library of Mrs. J. Montgomery Sears, Boston.

The Latin Herbarius.

1484. Herbarius Maguntie impressus. Anno ⁊ CLXXXIV. Printed at Mainz by Peter Schöffer.
(This is the book sometimes spoken of as Aggregator, but this word was never used as the actual title in any edition. The work is a Compilation from mediæval writers and consists of homely herbal remedies. The figures of plants are pleasing and decorative. The copy in the British Museum is not perfect, but there is a perfect copy in the Kew Library.
America: Missouri Botanical Garden, St. Louis.
1485. Herbarius Pataviæ impressus Anno domī ⁊ cetera LXXXV. Printed at Passau by Conrad Stahel.
(This edition is sometimes known as Aggregator Patavinus.)
America: John Crerar Library, Chicago.
1486. Another edition printed at Passau.
Undated editions. There are several in the British Museum. It is believed that one of them belonged to Sir Thomas More.
America: J. P. Morgan Library, New York.
1484. (*Flemish translation.*) Flemish translation printed by John Veldener Kiulenborg.
1500. Edition evidently a reprint of above printed by W. Osterman at Antwerp.
America: Hawkins Collection, Annmary Brown Memorial, Providence.
(*Italian editions.*)
1491. Edition printed at Vicenza by Leonard of Basel and William of Pavia.
America: Boston Medical Library, Boston.
1499. Edition printed at Venice by Simon of Pavia.
America: Surgeon-General's Library, Washington.
1502. Edition printed at Venice by Christ. de Pensa.

1509. Edition printed at Venice by W. Rubeum et Bernardinum Fratres Vercellenses.

1534. (*Italian translation.*) Herbolario Volgare nel quale le virtu de la herbe, etc., con alcune belle aggionte noua mētē de latino in Volgare tradulto. Printed at Venice.

Subsequent editions, 1536, 1539, 1540.

(In the Italian editions and translations the book is erroneously attributed to Arnold de Nova Villa, whose name is mentioned on the title-page with that of Avicenna. This error is pointed out in the British Museum Catalogue.)

1485 (*circ.*) (*French edition.*) Printed at Paris by Jean Bonhomme.

Herbarius zu teutsch.

1485. Herbarius zu Teutsch. Printed at Mainz by Peter Schöffer.

America: Surgeon-General's Library, Washington, and library of Mrs. Montgomery Sears, Boston.

The illustrations in this herbal are evidently drawn from nature, and are generally held to be only surpassed by those in the herbals of Brunfels and Fuchs. The preface is singularly beautiful. Though the preface enjoins the name "Ortus Sanitatis, in German, a Garden of Health," the title in this and subsequent editions is Herbarius zu teutsch.

1485 (a few months later than the above). Pirated edition printed at Augsburg, probably by Schönsperger. It is interesting to note that in this edition a pine cone, the badge of Augsburg, appears on the title-page. Figures of plants are very inferior to those in the first edition.

1486. Edition printed at Augsburg by Schönsperger.

Subsequent editions, 1487 (?), 1488, 1493, 1496, 1499, 1502. There are several undated editions.

America: Copy of edition printed in 1493 in Library of the College of Physicians, Philadelphia.

Arbolayre.

1485 (*circ.*) Arbolayre contenāt la qualitey et virtus proprietey des herbes gômes et simēces extraite de plusiers tratiers de medicine cōment davicene de rasis de constatin de ysaac et plateaire selon le coñu usaige bien correct.

(Supposed to have been printed by M. Husz at Lyons. It is believed that there are now only two copies of this book extant. One is in the Bibliothèque Nationale, Paris. The other was sold in London, March 23, 1898.)

Le Grand Herbier.

Before 1526. Le Grand Herbier en Francoys, contenant les qualites vertus et proprietes des herbes, arbres gommes. Printed at Paris by Pierre Sergent.

Ortus Sanitatis.

1491. Ortus Sanitatis. Printed at Mainz by Jacob Meydenbach.
(This is often regarded as a Latin translation of the Herbarius zu teutsch, but it is much larger and owes very little to that work. The woodcuts are copied from the Herbarius zu teutsch, but they are inferior.)
America: Surgeon-General's Library, Washington; John Crerar Library, Chicago; Arnold Arboretum, Boston; Mrs. J. Montgomery Sears Library, Boston; and J. P. Morgan Library, New York.

1511. Edition printed at Venice.
There are several undated editions.
America: Library of Congress, Washington; Arnold Arboretum, Boston; Surgeon-General's Library, Washington; Dr. G. F. Kunz's Library, New York.

1500 (*circ.*) (*French translation.*) Ortus Sanitatis translaté de Latin en francois. Printed at Paris by A. Vérard.
(The copy in the British Museum belonged to Henry VII.)

1539 (?) Edition printed at Paris by Philippe le Noir with the title "Le Jardin de Sante translate de latin en francoys nouvellement Imprime a Paris. On les vend a Paris en la rue sainct Jacques a lenseigne de la Rose blanche couronnee. (*Colophon*) Imprime a Paris par Philippe le noir."

Æmilius Macer.

1491 (?). Macer floridus De viribus herbarum. Printed at Paris.

1500 (?). Another edition. (Paris?)

1506. Herbarum varias q̃ vis cognoscere vires Macer adest: disce quo duca doct'eris. (*Colophon*) Impressus Parisius per magistrum Johannem Ieune. Pro Magistro Petro Bacquelier. 1506.

1588. (*French translation.*) Les fleurs du livre des vertus des herbes, composé jadis en vers Latins par Macer Floride. Le tout mis en François par M. Lucas Tremblay, Parisien . . . Rouen.

Jerome of Brunswick.

1500. Liber de arte distillandi de Simplicibus. Johannes Grüeninger, Strassburg. 1500.
(*English translation.*) See Bibliography of English Herbals.

Johann Czerny.

1517. Kineha lekarska kteraz slowe herbarz. Hieronymous Höltzel. Nürnberg. 1517.

Otto von Brunfels.

1530. Herbarum vivæ eicones ad nature imitationem, sum̄a cum diligentia et artificis effigiate. . . . Argentorati apud Ioannem Schottum.
Subsequent editions, 1530, 1531, 1532, 1536, 1537.
(The illustrations in this herbal are much superior to the text, which is based chiefly on the writings of the Italian herbalists. Brunfels was a Carthusian monk who turned Protestant. Jacob Theodor (Tabernæmontanus) was a pupil of Brunfels.)

Eucharius Rhodion.

1533. Kreutterbůch von allem Erdtgewachs Anfenglich von Doctor Johan Cuba zusamen bracht Jetzt widerum new Corrigert. . . . Mit warer Abconterfeitung aller Kreuter. . . . Zu Franckfurt am Meyn, Bei Christian Egenolph. 1533.
(The above was not an original work, but merely a revised and improved edition of the German Herbarius. The illustrations are copies of those in Brunfels's herbal.)

Iean Ruel.

1536. De Nature stirpium libri tres, Ioanne Ruellio authore. . . . Parisiis Ex officina Simonis Colinæi. 1536.
(Jean Ruel was a physician and a professor in the University of Paris.)

Leonhard Fuchs.

1542. De historia stirpium effectis & expressis Leonharto Fuchsio. . . . Basileæ, in officina Isingriniana. Anno Christi 1542.
Subsequent editions, 1546, 1547, 1549, 1551, 1555.
1543. (*German translation.*) New Kreüterbůch. . . . Bedruckt zu Basell durch Michael Isingrin.
1557. (*Spanish translation.*) Historia de yeruas y plantas de Leonardo Fuchsio. . . . En Anvers por los herederos de Arnald Byrcman.

Conrad Gesner.

1542. Catalogus plantarum. . . . Authore Conrado Gesnero. . . . Tiguri apud Christoph Froschouerum.
(Gesner's most important work—a general history of plants—was never published, for he died of plague before it was finished. The illustrations he had collected were published by Christopher Jacob Trew 150 years later.)

Hieronymus Bock.

1546. Kreuter Bůch. Darin Underscheid Würckung und Namen der Kreuter so in Deutschen Landen Wachsen. Wendel Rihel. Strasburg.
Subsequent editions, 1539, 1560, 1572, 1577, 1595, 1630.
(The first edition (1539) has no illustrations. The illustrations in the second edition (1546) are generally supposed to have been copied from Fuch's Herbal (1542), but many of them are original. Bock's Herbal is remarkable for the accurate descriptions of the plants.)

Rembert Dodoens.

1554. Kruydeboeck. . . . Rembert Dodoens Medecijn van der stadt van Mechelen. Ghedruckt Tantwerpen by Jan vander Loe.
Subsequent editions, 1563, 1603, 1608, 1618.
1557. (*French translation.*) Histoire des plantes. . . . Nouvellement traduite de bas Aleman en François par Charles de l'Escluse. En Anvers De l'Imprimerie de Jean Loë.
1578. (*English translation.*) See Henry Lyte in Bibliography of English Herbals.
1566. Frumentorum, leguminum, palustrium et aquatilium herbarum aceorum quæ eo pertinent, historia. . . . Antverpiæ Ex officina Christophori Plantini.
Second edition, 1569.
1568. Florum et Coronarium odoratarumque nonnullarum herbarum historia. . . . Antverpiæ Ex officina Christophori Plantini.
1583. Remberti Dodonæi mechliniensis medici Cæsarei. Stirpium historiæ pemptades sex sive libri xxx. . . . Antverpiæ Ex officina Christophori Plantini.
Second edition, 1616.

Pierandrea Mattioli.

1563. Neuw Kreüterbuch . . . von dem Hochgelerten und weitberümbten Herrn Doctor Petro Andrea Matthiolo. . . . Gedruckt zu Prag durch Georgen Delantrich von Auentin.
Subsequent editions ("gemehret unnd verfertigt Durch Joachimum Camerarium"), 1590, 1600.

Antoine Mizauld.

1565. Alexikepus, seu auxiliaris hortus. . . . Lutetiæ Apud Federicum Morellum.
1575. (*German translation.*) Artztgartem . . . neuwlich verteutschet durch Georgen Benisch von Bartfeld . . . zu Basel bey Peter Perna.)

Nicolas Monardes.

1569. Dos libros, el veno que trat a de tod as las cosas que traen de nuestras Indias Occidentales. . . . Impressos en Sevilla en casa de Hernando Diaz, en la calle de la Sierpe.

1571. Segunda parte del libro de las cosas que se traen de nuestras Indias Occidentales. Sevilla en casa Alonso Escriuano, Impressor.

1574. Primeray segunda y tercera partes de la Historia medicinal de las cosas que se traen de nuestras Indias Occidentales en Medicina. . . . En Sevilla. En casa de Alonso Escriuano.
Second edition, 1580.

1574. (*Latin translation.*) De simplicibus medicamentis ex occidentali India delatis quorum in medicina usus est. . . . Interprete Carolo Clusio Atrebate. Antverpiæ Ex officina Christophori Plantini.
Subsequent editions, 1579, 1582, 1605.

1576. (*Italian translation.*) Due Libri Dell' Historia de I Semplici, Aromati, et altre cose, che Vengono portate dall' Indie Orientali, di Don Garzia Dall' Horto . . . et due Altri libri parimente di quelle che si portano dall' Indie Occidentali, di Nicolo Monardes, Hora tutti tradotti dalle loro lingue nella nostra Italiana da M. Annibale Briganti. . . . In Venetia.
Subsequent editions, 1582, 1589, 1605, 1616.

1600. (*Flemish translation.*) Beschriivinge van het heerlijcke ende vermaerde Kruydt wassende in de West Indien aldaer ghenaemt Picielt, ende by den Spaenaerden Tabaco, en van desselvē wonderlijcke operatien eñ Krachtengemaert by D. Monardes Medecijn dez stede Sivillen eñ overgheset Door Nicolaes Iansz vander Woudt. Tot Rotterdam, By Jan van Waesberghe.
(The title-page has a charming illustration of a little Indian boy smoking a long carved pipe, and a figure of the tobacco-plant.)

1619. (*French translation.*) Histoire des Drogues. . . . La seconde composée de deux liures de maistre Nicolas Monard, traictant de ce qui nous est apporté de l'Amerique. Le tout fidellement translaté en François par Antoine Colin, maistre Apoticaire juré de la ville de Lyon. . . . A Lyon, au despens de Iean Pillehotte, à l'enseigne du nom de Iesus.
(*English translation.*) See John Frampton in Bibliography of English Herbals.

1895. (*German translation of the* 1579 *Edition.*) Die Schrift des Monardes über die Arzneimittel Americas nach der lateinischen Übertragung des Clusius aus dem Jahre 1579. Übersetzt und erläubert von Kurt Stünzner, Dr. med. Mit einem Vorwort von Prof. Dr. Erich Harnack in Halle a S.

Bombast von Hohenheim (Paracelsus).

1570. [Ettliche Tractatus des hocherfarnen unnd berumbtesten Philippi Theophrasti Paracelsi. . . . I. Von Natürlichen dingen. II. Beschreibung etlichen Kreütter. III. Von Metallen. IV. Von Mineralen. V. Von Edlen Gesteinen. Strassburg. Christian Müllers Erben.]
(The " doctrine of signatures " is usually associated with the name of Paracelsus, but the greatest exponent of this theory was Giambattista Porta.)

Nicolaus Winckler.

1571. Chronica herbarum, florum seminum. . . . Authore Nicolao Wincklero, Forchemio, Medico Halæ. . . . Augustæ Vindelicorum in officina Typographica Michaëlis Mangeri.
(The above is an astrological calendar giving the times when herbs should be gathered.)

Bartholomaus Carrichter.

1575. Kreutterbůch des edlen vñ Hochgelehrten Herzen Doctoris Bartholomei Carrichters von Reckingen. . . . Gedruckt zů Strassburg an Kornmarck bey Christian Müller.
Subsequent editions, 1577, 1589, 1597, 1615, 1618, 1619, 1625, 1652, 1673, 1739.
(In this Herbal every plant is assigned to one of the signs of the zodiac.)

Charles de l'Escluse.

1576. Caroli Clusii atrebat. Rariorum aliquot stirpium per Hispanias obseruatarum historia, Libris duobus expressa. . . . Antverpiæ, Ex officina Christophori Plantini.
1583. Caroli Clusii atrebatis. Rariorum aliquot Stirpium, per Pannoniam, Austriam & vicinas quasdam Prouincias obseruatarum Historia, Quatuor Libris Expressa : . . . Antverpiæ, Ex officina Christophori Plantini.
1601. Caroli Clusii Atrebatis. . . . Rariorum Plantarum Historica. . . . Antverpiæ Ex officina Plantiniana Apud Joannem Moretum.
(A republication of the two works cited above with some additional matter.)
For De simplicibus medicamentes ex occidentali India, see N. Monardes.

Mathias de L'Obel.

1576. Plantarum seu stirpium icones. Antverpiæ Ex officina Christophori Plantini.
1581. (*Flemish translation.*) Kruydtboeck. . . . Deur Matthias de L'Obel Medecyn der Princ' exc^{en}. T'Antwerpen. By Christoffel Plantyn.
(The Flemish translation is dedicated to William of Orange and the Burgomasters of Antwerp.)

Leonhardt Thurneisser zum Thurn.

1578. Historia sive descriptio plantarum. . . . Berlini Excudebat Michael Hentzke.
Second edition, 1587. . . . Coloniæ Agrippinæ apud Joannem Gymnicum.

1578. (*German translation.*) Historia unnd Beschreibung Influentischer Elementischer und Natürlicher Wirckungen, Aller fremden unnd Heimischen Erdgewechssen . . . Gedruckt zu Berlin, bey Michael Hentzsken.
(Thurneisser was one of the foremost exponents of astrological botany. He gives astrological diagrams showing when the various herbs should be picked. The illustrations are not particularly good, but they are attractive owing to the quaint ornamental border which surrounds each figure.)

Andrea Cesalpino.

1583. De plantis libri xvi. Florentiæ Apud Georgium Marescottum.

Geofroy Linocier.

1584. L'histoire des plantes traduicte de latin en françois : . . . à Paris Chez Charles Macé.
(The above is based chiefly on the works of Fuchs and Mattioli.)

Castor Durante.

1585. Herbario nuovo di Castore Durante, medico et cittadino romano. . . . In Roma. Per Iacomo Bericchia & Iacomo Tormerij.
Subsequent editions, 1602, 1617, 1636, 1667, 1684.

1609. Hortulis Sanitatis. Das ist ein heylsam[es] vnd nützliches Gährtlin der Gesundheit. . . . Erstlich von Castore Durante . . . in Italianischer Sprach verfertigt. Nunmehr aber in unsere hoch Teutsche Sprach versetzt, Durch Petrum Uffenbachium Getruckt zu Franckfort am Mayn durch Nicolaum Hoffman.
(It is uncertain whether this is a translation of Herbario nuovo. See Meyer, *Gesch.*, IV. p. 383.)

Jacques d'Aléchamps.

1586–1587. Historia generalis plantarum . . . Lugduni, apud Gulielmum Rovillium.

Joachim Camerarius.

1588. Hortus medicus et philosophicus. . . . Francofurte ad Mœnum.

Giambattista Porta.

1588. Phytognomonica. . . . Neapoli Apud Horatium Saluianum.

Jacob Theodor (Tabernæmontanus).

1588. Neuw Kreuterbuch. . . . [Nicolaus Bassæus] Franckfurt am Mayn.
1590. Eicones plantarum seu stirpium. Nicolaus Bassæus, Francofurte ad Mœnum.
1613. Neuw Vollkommentlich Kreuterbuch . . . gemehret Durch Casparum Bauhinum. . . . Franckfurt am Mayn, Durch Nicolaum Hoffman. In verlegung Johannis Bassæi und Johann Dreutels.
Subsequent editions, 1625, 1664, 1687, 1731.

Fabio Colonna.

1592. ΦΥΤΟΒΑΣΑΝΟC sive plantarum aliquot historia. . . . Ex officina Horatii Saluiana. Neapoli. Apud Io Jacobum Carlinum & Antonium Pacem.
(This Herbal is the first in which copper-plate etchings were used as illustrations.)

Adam Zaluziansky von Zaluzian.

1592. Methodi herbariæ, libri tres. Pragæ, in officina Georgii Dacziceni.

Gaspard Bauhin.

1596. ΦΥΤΟΠΙΝΑΞ seu enumeratio plantarum ab Herbarijs nostro seculo descriptarum, cum earum differentijs. . . . Basileæ per Sebastianum Henric petri.
1601. Animadversiones in historiam generalem plantarum Lugduni editam. . . . Francoforti Excudebat Melchior Hartmann, Impensis Nicolai Bassæi.
1620. ΠΡΟΔΡΟΜΟΣ Theatri Botanici. . . . Francofurti ad Mœnum, Typis Pauli Jacobi, impensis Ioannis Treudelii.
Second edition, 1671.
1623. ΠΙΝΑΞ Theatri Botanici. . . . Basileæ Helvet. Sumptibus et typis Ludovici Regis.
1658. Caspari Bauhini . . . Theatri Botanici sive Historiæ Plantarum. . . . Liber primus editus opera et cura Io Casp Bauhini. Basileæ. Apud Joannem König.

Claude Duret.

1605. Histoire admirable des plantes et herbes esmeruillables & miraculeuses en nature. . . . A Paris Chez, Nicolas Buon demeurant au Mont S. Hylaire à l'image S. Claude.

Jean Bauhin and J. H. Cherlerus.

1619. J. B. . . . et J. H. C. . . . historiæ plantarum generalis . . . prodromus . . . Ebroduni, Ex Typographia Societatis Caldorianæ.
1650–51. Historia plantarum universalis. Auctoribus Johanne Bauhino, Archiatro. Joh. Henrico Cherlero Doctore : Basiliensibus Quam recensuit et auxit Dominicus Chabræus. D. Genevensis Juris vero publici fecit. Fr. Lud. a Graffenried. . . . Ebroduni.

Johann Poppe.

1625. Kräuter Buch . . . Leipzig, In Verlegung Zachariæ Schürers und Matthiæ Götzen.

Guy de la Brosse.

1628. De la nature, vertu et utilité des plantes. Divisé en cinq livres. . . . Par Guy de la Brosse, Conseiler & Medecin ordinaire du Roy. A Paris Ches Rollin Barragnes, au second pillier de la grand' Salle du Pallais.
(This Herbal is dedicated to Cardinal Richelieu. It is the only Herbal with mottoes on the title-page—" Chasque chose a son ciel et ses astres " ; " En vain la medicine sans les plantes " ; " De l'Esperance la connaisance.")

Antonio Donati.

1631. Trallato de semplici . . . in Venetia . . . Appresso Pietro Maria Bertano.

Petrus Nylandt.

1670. De Nederlandtse Herbarius of Kruydt-Boeck . . . t'Amsterdam Voor Marcus Doornick.
(The original drawing for the frontispiece by G.v.d. Eeckhout is in the Print room of the British Museum.)
1678. Neues Medicinalisches Kräuterbuch . . . Osnabrück bey Joh. Georg Schwandern.

INDEX